Ben Collins, formerly The Stig from the BBC's internationally acclaimed *Top Gear*, was the benchmark of speed against which hundreds of celebrities set themselves: the man that everyone – including Formula One stars Nigel Mansell and Lewis Hamilton – tried and failed to beat. He is also the go-to guy for Hollywood car chases (he was Bond's stunt driver in every film since *Quantum of Solace*) and has raced successfully in almost every class imaginable, from Touring Cars and Le Mans 24 Hour to the American Stock Car circuit.

Also by Ben Collins

The Man in the White Suit
How to Drive

ASTON MARTIN

Made in Britain

BEN COLLINS

Quercus

First published in Great Britain in 2020 by Quercus Editions Ltd

This paperback published in 2021 by

Quercus Editions Ltd
Carmelite House
50 Victoria Embankment
London EC4Y 0DZ

An Hachette UK company

A CIP catalogue record for this book is available
from the British Library

ISBN 978 1 52941 081 5
Ebook ISBN 978 1 52941 079 2

Part-titles images: pp. 11, 117, 221, 255 © Aston Martin;
p. 31 © Motorsport Images; pp. 67, 89, 185 © Aston Martin Heritage Trust

Plate section: p. 1 (top and bottom) © Motorsport Images; p. 2 (top) © Aston Martin Heritage Trust,
(bottom) © Press Association Images; p. 3 (top and bottom) © Aston Martin Heritage Trust;
p. 4 (top) © The Klemantaski Collection, (bottom) © Getty Images;
p. 5 (top) © The Klemantaski Collection, (bottom) © Aston Martin Heritage Trust;
p. 6 (top) © Aston Martin Racing, (bottom) © Jordan Pay; p. 7 (top) © Motorsport Images,
(bottom) © Getty Images; p. 8 (top) © DHL, (middle) © Aston Martin Heritage Trust,
(bottom) © Aston Martin

10 9 8 7 6 5 4 3 2

Typeset by CC Book Production
Printed and bound in Great Britain by Clays Ltd, Elcograf S.p.A.

MIX
Paper from
responsible sources
FSC® C104740

Papers used by Quercus Editions Ltd are from well-managed forests and other responsible sources.

For Dad,
and all those who love a little freedom under their right foot.

CONTENTS

INTRODUCTION

Motor racing is about people, not cars.

Sammy Davis, Aston Martin works
driver and record breaker

A ston Martin is like a high-speed baton that never slows down, always passing from one set of loving hands to the next. Its raw energy is the continuous pursuit of perfection and this defines its soul. That's not to say the cars are all perfect – they made the occasional poltergeist too – but the ethos of ultimate performance is undeniably there and it has attracted successive generations of brilliant people to the banner. This book is about those people and some of their achievements, with the occasional interlude where I manage to get my hands on the merchandise.

The stories do go off on tangents, but that was the only way to enjoy the butterfly effect of these endeavours – most notably with Sir Stirling Moss, who personified perfectionism; a force of nature who remained a mystery to me until I really dug into his method. He really was *without equal*.

I received a lot of help from the wonderful folks at 'Astons', who helped me to shine a light on the fascinating parts of their history and agreed to be interviewed to add life to the tale. I grew to admire the characters I was unable to speak with, and at times have taken the liberty of giving them a voice in the occasional passage where I felt the story needed it.

As any engineer I've worked with would gladly confirm, I'm not a nuts and bolts man. I leave that to the professionals. To better understand what constitutes success and failure in the field of engineering, however, I have immersed myself in some outstanding books on the subject. For those of a mechanical disposition, I've listed the real belters in the bibliography.

I've always been fascinated by the individuality that steers design and I hope you will enjoy the way this story weaves its way through a hundred years of pioneering development and raw speed, oil slicks and ejector seats, bombs and bullets. The first century of Aston Martin contains its fair share of thrills, spills and crushing defeats, but ultimately it is a very British story of a never-say-die spirit and a constant drive to be, quite simply, the best.

PROLOGUE

EXPENSIVE SOCKS

1993, the kitchen table

Whoever said that working from home was fun or easy had obviously never tried it. My A-level exams were looming. I was groaning through English comprehension, spied my mum about to throw some of my revision notes into the bin and managed to intercept them. I grudgingly re-engaged with Thomas Hardy and cursed his observations on the Industrial Revolution.

We lived in a Devon longhouse that was about 400 years old, where each room connected to the next in line. The walls were four feet thick in places and the original dwellers not much taller, judging by the ceiling heights. The familiar thump of heavy feet from the dining room meant that Dad was taking a break from his office in the barn next door. He entered the kitchen looking purposeful, wearing a sports jacket and a raised eyebrow.

'Seen my keys?'

'Inside the chicken.' I waved vaguely in the direction of the ornamental pot on the worktop.

'I'm just popping out to buy some socks.' Then, looking at me, 'Want to come?'

Besides Thomas Hardy, the only thing to look forward to was racing the farm's 250 cc quad bike around the outside of the house and over the fields. Unfortunately, since I'd balded the tyres on previous forays, Mum had taken precautions: she'd emptied the fuel tank and hidden the keys. She was thorough, I'll give her that. Clearly, therefore, Dad was posing a rhetorical question.

We climbed aboard the company car, a Mercedes that belched *eau de vindaloo* from the air con until it had been running for a while. We clipped along the tight, overgrown Devon lanes and out of the village towards the 'sock shop' . . .

Devon's lanes are wonderful for driving. Dusty ribbons of constantly unfurling corners, diving down between pastures and the occasional short shoot through a copse. Dad went at it with aplomb, confident of the stopping distances required for a chance encounter with a hard-charging milk tanker and taking every opportunity to put the pedal to the metal. The Merc was putty in his hands and quickly filled with smoke from a steady chain of Marlboro 100s. The windows could only be opened when he stubbed one out because the draught blew ash onto the seats and he didn't like that.

We cruised through the bustling market town of Crediton, home to Blockbuster and Gateway and to my mind plenty of worthy sock-selling establishments. But we left town and he opened the tap on the Merc. Next stop, the bright lights of Exeter.

Avoiding the High Street, we raced on to Marsh Barton trading estate, where the colourful flags of various car manufacturers fluttered invitingly, beckoning us into a world of gaudy discounts stamped on windscreens and sales reps eager to clamp your cash between their clammy fingers.

We passed the Honda dealership. Dad had bought an NSX in 1990 and the handling was next level. Having perfected their high-revving, unbreakable F1 engines with a little help from Ayrton Senna, Honda had put the experience to work in their first ever supercar. By placing the V6 power plant in the middle like a Ferrari, they created a lightweight sports car that you could grab by the scruff of the neck and really hurl into corners. It wasn't a machine for the tender-footed, because at low revs it felt like a Civic. You had to stretch it out past 6000 rpm for the variable-valve timing to really open up the lungs and deliver the power. But . . . it still lacked something.

We pulled into a dealership that seemed to sell everything from Bentley to Maserati, 'just to take a look'. We poked round a Range Rover for a bit, admiring its regal good looks. A sales rep appeared out of thin air and exchanged the usual pleasantries.

What were we looking for?

Socks, I thought.

The rep sized Dad up from his dapper brogues to his designer sweater and ushered us into a covered area. The honey trap was well and truly primed with a feast of low-slung, brooding machines and sculpted metal. We kept a steady pace, until . . . *BRAAPPPP-BRAAAAAP*. . .CH-CHUNG-CHUNG-CHUNG-CHUNG-CHUNG . . .

The sound waves shook my internals. I felt like I was five years old again, at my first air show, watching a sinister triangular-shaped aircraft heave into view as the pilot yanked the stick and slammed open the afterburners, air crackling as I instinctively tried to burrow into the ground. That was the Vulcan bomber. This was another level, and a whole lot nearer.

I stopped and turned towards the source of the deep, gurgling rumble to my right. Two piercing red eyes shone back from the tail lights of the beast lurking in the half-light.

'What's *that*?' I asked.

'The Aston Martin Vantage.' The rep watched with almost indecent satisfaction as we swallowed his bait, hook, line and sinker.

We moved slowly towards it.

The wheels beneath its sinuous shoulders were colossal. As we got closer, the colour of the bodywork revealed itself as a deep, luxuriant British racing green. We climbed aboard and in contrast to the dark exterior, the cabin was flush with white leather and green trim. The seats were exquisitely comfortable, cushioned yet supportive. A large herd of cattle had undoubtedly sacrificed themselves to make them so.

Man-sized handles abounded. The handbrake looked like a truncheon, the gearstick more of a cleaver. The vast green dashboard curved around a driver console at the centre of which was the helm: a magnificently proportioned wheel with a similarly imposing centre frame. A parp of the horn on a device of this magnitude would scatter small animals within a one-mile radius. Even the key was a brick.

The rep leaned towards my old man and began his patter. I almost pitied him. Dad was one of those people who could gauge someone's worth, or, in his words, whether they could 'walk and chew gum at the same time', before they even entered his office.

The rep explained the limited production run and the high specification of this particular specimen. This was the only car not yet reserved from their exclusive quota, but interest was strong; it wouldn't sit on the forecourt long . . .

My dad had negotiated terms with trade unions, won international distribution tenders, merged and floated companies and was once even presented with a signed plaque by the President of the United States for his outstanding contribution to the trucking industry. As the MD of a multi-faceted FTSE 100 enterprise, he could eat this guy for breakfast.

I saw his knuckles whiten as he gripped the wheel a little more tightly.

'I'll take it,' he said.

The corporate man had left the building, replaced by the spirit of the teenager who quit grammar school with a handful of O-levels, left home and grafted from rock bottom in pursuit of a dream. A dream manifest there and then in his hands.

I looked across at him with a confused mix of surprise and glee. He took the car for a perfunctory spin with the rep. I wasn't fussed about going along as I knew he wouldn't give it the beans until we had it to ourselves. Dad signed the paperwork and we were off. Presumably he had a plan for the company car we had abandoned in the car park, or more likely had forgotten all about it.

I was intrigued by the manual shift as we peeled out of the dealership. For starters, it had six forward gears, but we only seemed to need two to break the speed limit. We cruised through the centre of Exeter, exuding charisma. All eyes were drawn to the sullen expression of the radiator grille, the take-it-or-leave-it headlamps, the 5.3 litres of pulsating aggression that threatened to burst out of the bonnet at any moment.

The Vantage was the ultimate badass. It made Mad Max's interceptor look like a social worker's minivan. A seventeen-year-old lost in a haze of Lynx Oriental, I daydreamed that the magic of its lean lines and atomic rumble weren't lost on the fairer sex either, our throaty acceleration away from traffic lights echoed by a crescendo of snapping knicker elastic for miles around.

Leaving Exeter behind, Dad gunned the accelerator as we left a roundabout. The brawny suspension absorbed the load but the beast got angry. The pair of Eaton superchargers clamped to the enormous V8 engine had awoken, unleashing 550 lb of torque into the rear axle. Strong forearms corrected the resulting twitch.

'Strewth!' Dad giggled. He was turning into an Australian.

A moment later we careered over an old railway crossing and the revs leaped higher as the tyres momentarily broke free.

'*Jesus Christ!*' And a Christian too.

I gripped the door support and we sallied forth. The Vantage demanded your full attention, that was clear, and so too was the cabin: the old boy didn't spark up another cigarette until we parked the car in our driveway. We both let out a sigh of relief after the sheer excitement of the drive.

The Vantage was a source of wonder. How could such a car

come into being? I grew up watching James Bond movies like everyone else, but this machine was on a different plane from the little silver-birch-coloured DB5. You could bet your life that we would be slipping *Goldfinger* into the VHS that evening to look for answers, but before that happened, we had to get past Mum. We climbed out and swung the doors, which closed with a considerable *wallop*.

'Oh, Bill!' she exclaimed. *So much for buying socks*, was what she was thinking. He had only gone and bought another bloody supercar, although at least this one could have ploughed the fields. She approached slowly for a closer inspection, eyes twinkling with excitement.

PART ONE
1902–14

Coal Scuttle

THE ASCENT OF MAN: FROM TWO LEGS TO FOUR WHEELS

Need for Speed

At the tail end of the nineteenth century people were getting a taste for speed, and they wanted more. Horses could achieve a little over 40 mph but would tire over long distances, plus they had the tendency to bite. Combustion and steam engines were slowly replacing horsepower, while human *leg* power was being harnessed by the humble bicycle. But as the twentieth century dawned, a gifted generation of engineers and test drivers from a range of backgrounds were playing with fire, steam and electricity in their bids to capture lightning in a bottle.

It was from this primordial soup of boffins, racers, tinkerers and speed merchants that a British motoring icon was forged. The path of those early years may not always have been straight, as a constellation of new races and disciplines proliferated, alliances were made and broken and ideas exchanged, borrowed or even stolen to advance the cause. But it is only by seeing the world through the eyes of those trailblazers and tracing their myriad solutions to the challenges of the day that we can revel in their times and enjoy the end product as much as they did.

Today we take it for granted that a car has four wheels propelled by an engine. Imagine what it would be like to start with a blank canvas and invent a totally new mode of transport from scratch. Perhaps then it will not come as a surprise that one of Aston Martin's most potent and enduring engine designs was the outcome of a life's work in locomotives and aviation, or that the company's founder, the father of a legend on four wheels, began life as a pioneer on two.

I would no sooner balance my crown jewels on a penny-farthing than use an electric shaver in the bath. Like-minded Victorians waited for John Boyd Dunlop to invent a pneumatic tyre to replace the bone-jarring solid wheels of the day before committing to the bike in droves. In 1896 the *New York Times* heralded bicycles as a 'splendid extension of personal power and freedom, scarcely inferior to what wings would give'. To the Victorian schoolboy, these wobbly contraptions were a source of considerable amusement that would soon develop into a craze. In addition to requiring mastery of balance, there was the physical strength and endurance needed for long-distance riding as well as the unadulterated thrill of speed.

Lionel Martin was born in Cornwall in 1878 and arrived at Eton College when he was thirteen years old with a bike his mother had given him. His robust frame adapted naturally to the rigours of public-school life and the challenges of his two-wheeler. By the time he reached Oxford University he was in peak physical condition. One perk at Oxford was a cinder track around which cyclists could compete at breakneck speed. Lionel soon proved himself a winner, both against Cambridge and notably in the longer-distance competitions organized by the Amateur Cycling Association.

In 1902 he was invited to join the exclusive Bath Road Club. This was the most select of all the road clubs with a membership limited to 130 high-rolling speed freaks. Bath Roaders got their kicks from extravagant dinner parties and gruelling road races, largely across southern England. Lionel more often than not led from the start, with occasional stops to feast on gargantuan quantities of protein and old ale. He burned with a desire to push his boundaries beyond the ordinary.

In 1902 he set off for London from Land's End in Cornwall in a bid to break the record by bike. Despite crashing within sight of the start line, Lionel pressed on to Bodmin, then Exeter and across to Devizes, where he spent the night. For this marathon he substituted gallons of ginger beer and milk for his usual ale. The next morning his tired legs pumped the pedals again. He blasted through Reading and reached London in time to destroy the previous record by over three hours.

The secret of Lionel's success lay not just in his physical and mental strength. He rode tactically, toying with competitors psychologically before breaking their spirit. And he was a tinkerer too. He meticulously fine-tuned every component on his machine to improve its reliability and performance. He polished every bearing, balanced the chain, optimized his saddle position and carried essential spares on his back. Bath Road racers flocked to him to give their rides the same treatment. Lionel also took with him a dapper change of attire for fine dining. He was ever the social dandy.

It was around this time that Lionel was struck by the next-generation power of the motor car, quite literally: 'I was cycling from my crammers (training) to London along the Oxford Road when I saw the monster approaching and I threw myself and "iron" into the nearest ditch.'

As Lionel lay in that ditch, he ruminated on the endless possibilities of the internal combustion engine.

Fast and Furious

Not everyone was enthusiastic about quitting their two-wheeled or four-legged friend for the noisy and smelly mechanoids invading the roads. The future war poet Siegfried Sassoon initially refused to move with the times, but once he did, his penchant for dangerous driving gave dangerous drivers a bad name. On his first outing he crashed into a dog cart, and then wiped out a cyclist on day two. He never signalled and would abruptly pull on to main roads regardless of oncoming traffic. It was enough to terrify the most fearless of passengers, including one T. E. Lawrence. Mechanically speaking, Sassoon fared little better. While inspecting his fuel tank with a lit match, he was admiring his reflection in the glassy liquid when the vapour ignited and blew his eyebrows off.

Perhaps understandably, in the early 1900s the public perception of motorists was mixed. One of Lionel Martin's Bath Road chums was the mega-moustachioed motoring pioneer and cycling champion Selwyn Edge. He was out taking the air in his motor one morning, waving as he passed the driver of a horse-drawn cab, when he felt the flick of leather about his ears. In perhaps one of the earliest recorded instances of road rage, he stopped to face the culprit, removed him from his cab and flogged him thoroughly with his own whip.

As a nation the Brits were broadly sceptical of motoring and passed a draconian law requiring a man with a red flag to walk in front of each vehicle to restrict its speed. This law was repealed in 1896, but the UK limit remained a pitiful 12 mph. The continent was already leading the technological and sporting development of the breed. France was in pole position, with hundreds of automobile manufacturers producing a variety of models, with Paris the epicentre of motor mania.

Motoring as a sport began with the Automobile Club de France organizing a series of tortuous races between Europe's capital cities, including a 700-miler from Paris to Berlin in 1901. Invading Germany for once made a pleasant, if brief, reversal of fortune for the French. The winner covered the distance at an average of 44 mph, which meant that drivers attained considerably higher speeds in the open sections.

Thousands of police patrolled the course in a bid to restrain the huge crowds of spectators from trying to get among the action. Blue-flag men vaguely signalled the more dangerous stretches. Before the race even began the throng of spectators and competitors moving through the night towards the start line was a sight to behold: 'the cyclists were carrying paper lanterns of all shapes, colours and sizes . . . a great multi-coloured serpentine of light winding away through the trees'. The 110 vehicles were grouped according to weight with everything from motorcycles and light cars weighing up to 880 lb, to the heavy brigade limited to 1450. The rules stated that a mechanic or chauffeur had to ride shotgun in the passenger seat, although they spent most of their time surfing the footboard alongside to balance the cars' weight.

Despite the crude power delivery to skinny tyres that wouldn't look out of place on a modern beach bicycle, the traction was reasonable and the tyres also provided good cornering bite. A bigger challenge was keeping the cars in a straight line as they wandered with the contours of the road surface – and slowing down with rear-only brakes.

The drivers starting the race had no idea what to expect round each bend or even how their cars might behave, since they were constantly under development and tuned to the maximum for these big events. Wheels flew off, engines blew out, stuff caught fire and brakes failed. It required Herculean strength to turn the steering wheels, and yet they threw these contraptions around with finesse and abandon. The French fielded the strongest contingent of men and machines, but among their ranks was a pair of plucky Brits.

Charles Jarrott was driving a French Panhard and pushing so hard that he was regularly up on two wheels, with his mechanic, Smits, leaning fully out from the footboard as a counterweight. As Jarrott sped towards the commune of Viels-Maison in northern

France, he checked over his shoulder. A giant moustache leered back at him through the clouds of dust, Jarrott finding 'to my horror that Edge, who had started a considerable time after me, had caught me and was just behind'. For a racing driver there is no more bitter pill to swallow than to believe that you're driving on the limit, only to discover that someone else has set the bar higher.

Jarrott opened the throttle fully in a bid to outrun Edge. He notched up 80 mph before cresting a rise beyond which lay a severe right-angled bend. He skidded into the corner, the car's spindly wheels wobbling across the loose surface. He made it round by the skin of his teeth, which was more than could be said for poor Smits, who, unprepared for the turn of events, was catapulted over a stone wall and came to rest next to the remains of another – wrecked – vehicle. Edge flew past Jarrott, but came a cropper himself after volleying his English-built Napier across a drainage gulley, crushing the suspension as he landed on the far side.

Leading from the front, the French ace Henri Fournier enjoyed a clean drive to the finish, where all the spoils of victory were lavished upon him. Yet the camaraderie was such among motoring fans that when Jarrott made it to the finish line in tenth place he too was showered with flowers, cigars, champagne and glory. Fournier dabbed the dirt from his mouth and claimed that the secret to his success lay in 'caution more than nerve. The reckless man will never win.'

Some would claim *au contraire* . . .

Gordon Bennett!

Gordon Bennett owned and ran the *New York Herald* and was as successful at creating the news as reporting it. Having bankrolled Stanley's journey in search of Dr Livingstone and established the sport of tennis in America, he created the world's first truly international motor race, the Gordon Bennett Cup, which was effectively the forerunner of the Grand Prix circuit. Each country was given the chance to enter a team, and the winning nation took home with them the right to host the following year's event. The French inevitably dominated the opening contests, until the unthinkable: in 1902, the Brits won. But it almost never happened.

The 1902 Gordon Bennett Cup ran concurrently with the Paris to Vienna road race, the difference being that the Cup race terminated short of Vienna at Innsbruck. Driving for Great Britain, Selwyn Edge was hammering a Napier Light in his customary pedal-to-the-metal style. His tyres were holding up well, against all expectations. However, the leakage of air from the continuous pummelling had left them flat, and having lost his tools and spares, he was stranded short of the Swiss border in desperate need of a pump.

Edge stood in the road and waved for assistance as a Mercedes 'heavy' tore into view. The driver pulled in. His jet-black hair and luxuriant moustache (which spread down to his chin) became increasingly animated as he spoke. Edge explained his predicament. The driver loosened his dustcoat, revealing an immaculate suit with a high-collared shirt and neatly knotted tie, and rooted about inside his toolbox. This prince among sportsmen handed Edge his pump and, without waiting for its return, set off with a roar. His name was Count Eliot Zborowski, a flourish of a surname that would resonate strongly through the early years of Aston Martin.

The next leg of the journey crossed Switzerland, where racing was forbidden. The competitors paced themselves while customs officials monitored arrival and departure times; in the event of a tie, the time spent transiting the country would be taken into account when it came to judging the overall winner. Then came a tortuous trail cut into the flanks of the Arlberg mountains. The vicious climbs put an extraordinary strain on oxygen-starved engines, forcing the drivers to work their clutches and gears to death. Just a few boundary stones separated the loose rocky surface from oblivion. Travelling downhill towards the tight hairpin bends frazzled brakes and nerves alike; many ploughed into the mountainsides as the only means of slowing down once their brakes had terminally overheated.

Only one entrant reached the Innsbruck finishing line, and it was Edge. While a protest was lodged against him for receiving assistance, the main race continued towards Vienna. French hero Marcel Renault was the first to come in and was declared the winner, despite having taken a short cut. Half an hour later, Zborowski thundered in aboard his Mercedes. The French officials, noting the German marque, added penalty time to his result for 'customs infringements' in Switzerland. For many, including Charles Jarrott, this was disgraceful: 'It is the opinion of some today that a mistake was made, and that in reality Zborowski was the winner.'*

Meanwhile, the protest against Edge was grudgingly dropped and he was declared the victor in the Gordon Bennett Cup. While the result shook the French, it galvanized the British automobile industry. The saying 'Win on Sunday, sell on Monday' had never been more true: sales of Napier Lights tripled off the back of this success.

---

* The following year, Eliot Zborowski took his big Mercedes to the French Riviera to run the gauntlet of tight turns and imposing rock to the pretty village of La Turbie, which overlooks Monaco. Charles Jarrott observed that Eliot always drove to win 'as if his life depended on it'. Rumour had it that prior to the race he had met with a fortune teller, and whether or not that was true, he signed his final will and testament the following morning, on April Fool's Day.

Eliot brought his 60-horsepower Mercedes to the line. If he felt any trace of nerves he didn't show it, suppressing dark thoughts by displaying every confidence that this was *his race*.

'But you know,' argued his rival Fernand Gabriel, 'all that matters in a race like this is to arrive alive at the finishing post.'

Eliot's foot quivered on the heavy clutch pedal as he waited for the start signal. The sun blazed on the long bonnet, and the engine temperature was rising. Eliot involuntarily released the clutch and lurched forward. A false start. He hauled the Merc back to the line and waited for the flag . . . to drop . . .

Second time round he used the full might of the engine, roaring away and up into fourth gear towards a hard right-hander. A light shower had formed a coating of greasy dust on the road surface, and as Eliot went to slow for the corner his starched, cuff-linked sleeve caught in the throttle advance of his steering wheel. Unable to reduce speed at the vital moment, he fought a losing battle with the wheel. The Mercedes thundered into the rock face, launching Eliot head-first into the wall and killing him instantly.

The Machine

In the wake of Napier-gate, Lionel Martin travelled to Paris for the motor and cycle show. He was impressed by the standard of the foreign models, less so by the British. He tried out a French Mors and ventured out in a friend's Mercedes, finding the local roads disappointing by home standards. What was needed was a proper British touring model that could flex its muscles across Europe.

Lionel went into business with yet another Bath Road cyclist, Robert Bamford, and the two of them traded and tuned a variety of small motors. The intimate knowledge they had gained across a range of machines soon became apparent. Lionel bought himself a little 10 hp Singer at its London launch in 1912 – a narrow, light and sturdy vehicle with a four-cylinder engine that could propel it to a less than hair-raising 40 mph. He took it back to their workshop in South Kensington and pulled it apart.

As a cyclist, Lionel could well appreciate the evolving sequence of pistons rising and falling inside the engine. The pistons were linked to a crankshaft that delivered power to the wheels much like pedals on a bike. A full sequence demanded two revolutions in order to suck the fuel and air in, squeeze it into the ignition chamber and explode it, and finally blow out the exhaust gases.

Average humans balance this rhythm of breathing and effort instinctively. Sportsmen hone it to an art form. In Lionel's day the timing of an engine had to be balanced mechanically by linking the power of the crankshaft to the breathing valves controlled by the camshaft. By modifying these whirling elements and testing their characteristics obsessively on the road, he figured out how to reduce friction and raise enough power in the Singer to take it to 70 mph. Lionel demonstrated his creation to good effect at the Aston Clinton hill-climb events in 1914, where he fell in love with the winding trail through Lord Rothschild's wooded estate even as the clouds of war began to gather over Europe.

Bamford and Martin became infuriated by the fact that they were constantly righting the wrongs of other manufacturers, which, happily for this book, left only one option: they decided to make their own car. Considering a new name for the business with his wife Kate, Lionel fortunately dismissed anything involving fish or fauna, and settled on the more pulse-quickening Aston Martin.

Using stock items from respected suppliers, they created the 'Coal Scuttle', a nickname that perhaps brings the pulse sharply back to resting rate again, but nonetheless it was the first Aston Martin of all.

Fighting Spirit

Walter shuffled towards his new school, Clifton College, and repeated in his mind what his older brother Horace had told him. He climbed the steps to the patchwork red and limestone Tait's Town building, and braced himself for mortal combat. Walter was known to his extended family as the Bun on account of his rotund figure and shadowy eyes. With his pressed shorts, stiff oversized collar and knobbly knees, he hardly looked the fighter. But he detested bullies, and they were all he could think about.

Horace's words again: 'A deuce [hell of a lot] depends on how you make out – *Give it all you've got!'*

Lunchtime arrived and the boys assembled. The Gothic vaulted ceiling of the dining hall towered high above them. They scoffed their food, cleared away the tables and benches and created a boxing ring. Walter was paired against a worryingly well-built opponent. But thanks to Horace, he had been training. He lunged forward and landed a whirlwind of blows. To everyone's surprise, not least his own, he was for once victorious.

Never overly physical, Walter Owen 'W. O.' Bentley may have been bodily present at Clifton College, battling his way through rugby and scuffles alike – but his mind was full of steam. In class he was painstaking and methodical in getting to the absolute nub of a problem; just never in the time frame set by his teachers. He remained stuck in the third form until his departure at the age of sixteen, with a final school report lamenting that 'he tries, but is very stupid'.

W. O. returned to his Yorkshire roots in 1905 and managed to secure an apprenticeship with the Great Northern Railway. Here his deliberate approach found its home in the precise and serious business of forging the components of a steam locomotive and preparing them for assembly. He also discovered how real working men took pride in their labour, and how 'the time and trouble they took to help you over a problem that was holding you up was sometimes overwhelming; it seemed almost a point of honour'.

Long hours of daily graft lay ahead. W. O. prepared connecting rods, freshly cast from molten metal in their sand moulds, chipping away the swarf and polishing them to an even tolerance.

Having earned the trust of his peers, he was finally allowed into the assembly shop, where the giant bearings, wheels and casings were strong-armed into one engine. Each clockwork masterpiece of iron, bolts and steel was a boyhood dream made manifest before him. His first journeys on the footplate of the speeding engine, feeding the flaming heart of the beast, feeling the rush of air and the roar of steam, must have been magical . . . yet the perennial question asked by all youngsters reared its head: what now?

The 5 a.m. bike journey to work taught W. O. two things: first that he detested waking early, and second that if his feet on the pedals had some assistance in making the commute, perhaps from a motor, it would free up more sleep time. This led to the seminal purchase of a 3 hp Quadrant motorcycle that instantly became the envy of his five older brothers. Two of them, Arthur and Horace, followed suit and so began a fraternal competition: who could travel furthest and fastest around the country, hooked by the lure of the open-ended destination? Arthur bagged the rump-tenderizing record from John O' Groats to Land's End, while W. O. won gold medals for London to Plymouth and back, as well as London to Land's End. So much for japes, but what about money?

W. O. landed a job working for the National Motor Cab Company in 1910. Here he unpicked every meter-cheating ruse that the ingenious cabbies could come up with to separate cash from the sticky fingers of their employers. W. O. had to wage a delicate war of wits without causing a walkout. It helped that he was able to tweak the carburettors to significantly reduce fuel consumption *and* managed to improve the reliability of the most abused gearboxes in London.

Tuning the fleet of cabs flushed W. O. with enough cash to buy a lightweight French sports car. Its handling and turn of speed impressed him. Shortly afterwards, brother Horace spotted an ad in a newspaper: a business selling French sports cars was coming under the hammer. The main failing of the enterprise was a total lack of promotion of its vehicles or any form of validation by trials events of the kind in which the Bentley brothers were engaged. W. O. was drawn to the Doriot, Flandrin & Parant motors in the company stable. The sporty four-seater model offered more performance to customers than anything else on the British market. So in 1912 he and Horace took the plunge and bought the company, bringing a new firm into being: Bentley and Bentley. (Do not adjust your set; you're still in the Aston Martin book and all will become clear, in time . . .)

Working with the French manufacturer, W. O. enhanced the power of the DFP's four-cylinder engine by increasing its compression. Now it was time to put it to the test . . .

These days hill climbs are something of a sideshow to mainstream motorsport events like Formula 1, Le Mans and NASCAR, and the relevance of racing to the car buyer is debatable. But in 1912 the correlation was clear: if your car could scrabble up a steep, muddy hill faster than anyone else's and without littering the course with parts, then folk bought what you brought.

W. O. eased up the clutch on his two-seater DFP and lowered his cap to shade the sun from his eyes. As he drove towards the Chiltern Hills he felt the warm glow of adrenaline filling his veins. The machine was chirping along beautifully and it filled him with confidence. It evaporated when he turned onto the bottom of the hill at Aston Clinton. The staging area for the dusty hill climb was teeming with swaggering characters and speed merchants, while the tree-lined route was bristling with spectators looking down onto their heroes entering the curves below.

A cacophony of magnificent engine notes reverberated around the banking as, one by one, the motors blasted up the hill. W. O. had plenty of time for his spirits to drop as he queued up for his turn and observed the record-breaking times that seasoned drivers were posting on the chalkboard. He inched his wheels up to the start line, drew up the handbrake and held the clutch.

As with his first fight, W. O. had been practising. At the off he let fly the handbrake and slipped the clutch simultaneously. His party trick, one that had to be timed perfectly, was a clutch-less shift from first to second gear. The advantage of maintaining full throttle on an uphill stage was huge because it preserved momentum. However, if the dog gears mismatched in the process, he risked breaking the teeth that meshed them together and an ignominious early retirement.

W. O. gripped the lever tightly, snapped his foot briefly off the throttle and plunged the gear home. The hardest part done, he relaxed into the flowing bends and crests that concealed nothing sinister, blowing across the finish line at full pelt in top gear. The world came back into focus. He even remembered that he had a passenger. His fiancée Leonie had been riding alongside and innocently wondered if they had set a record. It turned out that he had set several.

With a story to tell and sell to customers, this kind of result was a boon for business. It also cemented a closer relationship with

Bentley's friends in France. On a subsequent visit to Monsieur Doriot's office at DFP HQ they laid out the plans for the next generation of sports car. During the meeting W. O. noticed an odd paperweight on Doriot's desk, a little piston made of some kind of alloy that sat lightly in his fingers.

'Aluminium, you know,' Doriot said.

Ever the dog with a bone, the prospect of a lightweight-alloy piston flying up and down inside his new engine fuelled W. O.'s imagination. He set about a series of experiments mixing copper and aluminium, cast the alloys in his piston moulds, tested them and finally landed on the perfect formula: 12 per cent copper to 88 per cent aluminium. The next problem, and one that others had found insurmountable, was that the alloy was prone to excessive expansion at high temperatures and under the resulting impacts broke up. W. O. got round that by carefully making allowances within the cylinder, so the pistons were loose to begin with but tightened up as they got hotter.

The result was an engine that rattled like half a tin of biscuits from cold but was stronger, lighter, more powerful and better in every other way. In 1914 W. O. took the prototype to Brooklands, the mighty 2.75-mile autodrome that would become the crucible of British motor racing, and smashed every record from the flying half mile at 89.7 mph to the ten-lap average speed mark. The new car would be released to the public as the Speed Model at the British Motor Show. Everything was ready, but then war broke out.

PART TWO
1914–25

Bunny, with Clive Gallop

BUNNY,
THE GRAND PRIX CAR
AND THE RACE TO RACE

Fly Boys

The outbreak of World War I took on a festival mood. Young men flocked to the recruiting stations to sign up and have a crack at the Boche. Robert Bamford joined the Army Service Corps, while Lionel volunteered his driving services to the Royal Automobile Club and undoubtedly terrified a host of military VIPs on their trips in and out of London. All the machinery at Henniker Mews, the London birthplace of Aston Martin, was sold to the Sopwith Aviation Company, and the Coal Scuttle was tucked safely away. But far from bringing progress to a standstill, the war would instead prove to be the crucible in which decades of success for the nascent company would be forged: a time of technological innovation and metallurgical discoveries, certainly, but also of a very different alchemy, the coming together of the 'rat pack' of charismatic talents who drove Aston Martin to new heights.

Men like Sammy Davis, whose relatives had returned from the Boer War with bits blasted off, were more sanguine about the potential outcome of the conflict, but no less up for it. Sammy had a smooth, dark Pacino-like countenance, and a winning smile was never far from his lips. In his school days he had borrowed and wrecked a penny-farthing with Malcolm Campbell. The speed bug properly took hold when his father took him to the 1903 Paris to Madrid road race. He looked on in awe at the great noisy machines and from then on only imagined himself at the wheel.

They told him that twenty-seven was too old for the infantry, but as luck would have it he bumped into a recruiter for the Royal Naval Air Service and things were looking up until a friend convinced him to move across to the RNAS armoured-car division. By another turn of fortune this was the unit that had turned down W. O. Bentley.

Sammy departed for France, where his mount was basically a large car with a turret Frankensteined on top and metal plates running down the side. The rubber wheels were not bulletproof. Nevertheless, its role was to advance in front of the main line and attack enemy strongpoints with its Maxim machine gun. As the most conspicuous item lumbering around the battlefield, however, it was the proverbial bullet magnet, especially in built-up areas. Sammy, a wicked sketch artist, devised a solution. He painted a ruined house onto a giant canvas and used this to camouflage the wagon.

W. O. was still searching for his calling, and with two of his brothers already killed in the opening phase of the war, things developed a sense of personal urgency. He contacted a man called Wilfred Briggs who worked for the Admiralty. Briggs' role was to coordinate the civilian engineering manufacturers working on engine design for Britain's new air force. W. O. showed Briggs the alloy piston that had proved itself beyond doubt in his mind. *But if the design was that good*, Briggs thought, *why wasn't the entire industry already copying it?* On the other hand, he found Bentley's conviction and salesman's charm compelling and took him on. 'You're an officer now. Go to Gieves and get a uniform with two rings on it,' he ordered.

Bentley reported for duty and shared the secret of his fabulous alloy pistons with Rolls-Royce and Sunbeam, who duly incorporated it into their aero engines. Then he moved to the Gwynnes factory, whose engineers had had the foresight to begin work on the French-designed Clerget rotary aero engine just before the war. The original Clerget engine's cylinders extended individually from a central point like the spokes of a wheel. In the centre was a rotating crankshaft that harnessed the power of the pistons and distributed it to the propeller. The insane part of the design was that the whole engine spun round like a demented Catherine wheel. But its strength was its lightness. With very few supporting parts or cooling systems, it had an excellent power-to-weight ratio, which meant everything in flight. However, in combat operations over France, they had been prone to seizing for reasons unknown, leaving the pilots helpless when their aircraft burst into flames and fell from the sky. They likened each flight to suicide.

W. O. was assigned to a RNAS squadron and dispatched across the Channel. Diving for cover when none other than the Red Baron strafed the airfield, he introduced himself to Petty Officer 'Nobby' Clarke, who was lying next to him in the ditch. Nobby led him to the work sheds, where W. O. tinkered with one of the engines until he felt he had resolved the issue. Without further ado, he was sent up with the dawn patrol to test his work.

W. O. received a crash course on how to fire his rear-facing Lewis gun without shooting off his own tail, before the pilot taxied onto the grass strip and took off into the sun. As he pushed open the throttle, the torque from the prop twisted through the airframe, trying to pull the aircraft down and to the right. Experienced pilots used this right-turning power to outmanoeuvre the German Fokkers flying against them. The less able were often caught out making left turns, when the gyroscopic effect resisted the turn, lifted the nose and potentially stalled the aircraft into a spin.

A swarm of hornets swirled over the battle lines ahead of them. W. O. was convinced he saw an enemy fighter banking towards them but his pilot had other plans, bearing away and out towards the coastline. After a lengthy flight and a hard landing, W. O. unclenched his numb fingers from the Lewis gun and put on his best face for the ground crew that rushed up to greet them. News had travelled fast about the man with the plan to cure their flying death-traps. W. O. eagerly opened the engine to check the pistons and was horrified to see the blue metal that indicated imminent heat failure. 'From then on the appalling sense of responsibility hung over me for the rest of the war, the figure of a pilot killed by engine failure leaning over my shoulder.'

He returned to work doubly determined. The spinning cylinders on the Clerget were thinned as much as possible and fitted with narrow fins to draw away heat. The problem was that the leading edge cooled more than the trailing half of the cylinder, which was shielded from the air, causing uneven expansion. The rings wrapped around the pistons to seal their passage inside the cylinder were breaking apart, and when that happened, the engine did too. It couldn't be fixed without a complete redesign.

Briggs indulged his protégé and secreted W. O. in the Humber factory, a safe distance from the jealous scrutiny he'd suffered at Gwynnes, where he set about creating an aluminium cylinder with an iron sleeve shrink-fitted inside it. The iron, with half the expansion rate of aluminium at high temperatures, would hold its form while the highly conductive surrounding metal drew away the heat.

Naturally W. O. included his magic alloy pistons, and their lightness enabled him to extend their range of travel within the cylinder. This extra leg room and improved ignition raised the power output from 130 to 150 bhp. He also built a second proto-type, a bigger, more powerful version that boasted a thumping 200 bhp.

The Admiralty flatly refused to adopt the new design until the Bentley Rotary 1 – the BR1 – had been proved, a mistake that in hindsight possibly extended the war by another year. However, the plans for the BR1 eventually landed on the desk of newly pro-moted Admiralty inspector Sammy Davis. The enhanced power figures scared Sammy, who had nightmare visions of the engine's gyroscopic torque causing the airframe to spin helplessly around the propeller, but he gave his seal of approval.

The BR1 quickly replaced the deadly older engines in the field, with Clive Gallop, a pilot from the elite 56 Squadron, ensuring that the Royal Flying Corps adopted them too. The new Sopwith

Camels could power up to altitude faster than their opponents and scored more kills than any other Allied aircraft. They turned the balance of power by winning the struggle for air supremacy and even helped bring down the venerable Red Baron. Bentley's measured approach, his understanding of thermal dynamics and acute attention to detail were leading him down an enlightened path of efficient combustion.

Brooklands

In 1918 German stormtroopers had one last crack at the Allied lines in a spring offensive that was initially successful but finally repulsed. The armistice that formalized the cessation of hostilities in November was met with an eerie silence and disbelief in the lines. The constant sense of threat had gone, but its echo remained.

The troops' return unlocked a new energy that grew into nationwide euphoria, but it wasn't all roses for the victors. W. O., who had worked so tirelessly away from home, lost his wife to Spanish flu. Sammy Davis applied to resume his post as a motoring journalist and found himself being interviewed by someone who had never fought. After Davis explained that his years of service had culminated in the aero engine programme, the desk jockey idly remarked, 'Well, that's of *some* use to us . . .', as though the rest was not.

Meanwhile Bamford no longer felt the youthful allure of speed, and stepped down from his position at Aston Martin, leaving Lionel alone at the helm with only his Coal Scuttle for company. However, a group of mechanically inclined fire-eaters back from the war was beginning to flock together at a place of mutual understanding, a place where daring and ingenuity could be measured on a stopwatch. And that place was Brooklands.

Long-distance racing of the kind that had been permitted on public roads at the turn of the century was banned after a race to Madrid in 1903 resulted in eight fatalities on the first day. In response, the Automobile Club de France closed off looped sections of road and competitors raced round these circuits for a set distance. This model was adopted across Europe, and these events were called Grands Prix. Entrants had to adhere to regulations in order to qualify, but they were still just racing around on dusty roads.

In total contrast, and in a bold and unparalleled investment in British motoring infrastructure, wealthy landowners Hugh and Ethel Fortescue Locke King built the Brooklands Autodrome on a gargantuan scale in 1907 on their private estate in Surrey. It was the world's first purpose-built circuit, conceived and constructed to be a sort of motoring Ascot, and instantly became the Mecca of speed.

Hundreds of thousands of tons of earth were excavated and in just nine months a vast swathe of concrete traced the shape of a wonky pear. The left side was flat to create the long Railway Straight, which ran down into the expansive banked curve around Byfleet, returning up and inwards before forking right at the Fork, past the Vickers sheds, then uphill and left to swing round the top of the pear on the Home Banking, which ran underneath the Members' [only] Bridge.

The green-domed Clubhouse on the infield was the administrative hub as well as the weighbridge and a focal point for high society. In keeping with horse racing, the paddock was where the cars were berthed, and from there club members could scale the steps to the Clubhouse for a spot of luncheon, recline in the Ladies' Reading Room or shoot some billiards, observe the action from the gallery and then freshen up for further sport. The Members' Enclosure kept the hoi polloi out, and the guiding principle for admittance was to maintain the 'right crowd, and no crowding'.

At 2.75 miles in length Brooklands could have swallowed the Indianapolis Motor Speedway whole. The giant straights were 100 feet wide, half the width of a football pitch and twice that of Indy, with a dizzying 30-degree angle of banking that can only be matched today by the mighty Daytona. Vehicles roamed this sea of cement like so many ants. The circuit was designed for speeds not just of the time, but for all time.

Its inherent safety made it an ideal proving ground, and the likes of Selwyn Edge immediately set about posting records for average speeds over twenty-four hours, or from a standing start to a one-mile marker. Brooklands also lent itself to racing, and the track could accommodate an almost unlimited field. Crowds flocked to see the exotic machines that paraded under the watchful eyes of bookish officials.

An airstrip was added inside the circuit for good measure, and during the war years Brooklands had been home to a number of aircraft manufacturers including Sopwith, Hawker and Vickers. It was also used as a training facility for pilots before they headed overseas. The combination of aero and automotive engineering in a single facility of this scale would produce fascinating results.

Zborowski Returns

We last saw the dashing Eliot Zborowski making his mark in the Wild West of pre-World War I road racing, before the 1903 crash that cut him down in his prime.

The Zborowski family originated in Poland and had sailed to New York back in 1662 when it was still a Dutch trading outpost. Bartering with the Lenni Lenape tribe, they amassed landholdings that included a large part of Manhattan. Eliot's ancestors also included the man who penned the Constitution of the United States. His wife Margaret was a member of the Astor dynasty, which boasted similar wealth from fur trading with other native American tribes, and yet more property in New York. Their combined fortunes would pass into the hands of their son, Count Louis Zborowski, making him one of the richest men in the world at just sixteen years of age on the death of his mother in 1911. More relevant to our story, Lou, a future star in the Aston Martin firmament, was blessed with his father's razor-sharp racing genes.

When they arrived in England the Zborowskis were refreshingly American compared to the stiffer reaches of high society, and were renowned philanthropists. Eliot could outride the bravest horsemen and would jump anything. He was an early convert to autos, with young Lou riding on his lap from day one. The young count was sent to Eton, where he rebelled and returned to the family estate at Higham Park. The main house was of Palladian style and not dissimilar to the White House. During a refurb, Lou rode his motorbike around on the top of the scaffolding for a bet.

Higham had 23 bedrooms and was surrounded by 225 acres of parkland, not to mention several workshops. When not trawling London's nightclubs or exchanging ideas with the luminaries of Brooklands, Lou was to be found hammering away in the sheds. The scale of Brooklands had captured his imagination, as had the importance of horsepower in aeronautics. It seemed obvious that he should combine the two.

Lou extracted the chassis from a Mercedes truck and went shopping for his engine. Following the armistice, the War Disposals Board held a knockdown sale of motors from captured German Zeppelins and Louis bought himself a whopper. The Maybach MbIVa was designed to overcome the loss of horsepower associated with high-altitude flying. The thin air starved engines of oxygen and reduced power, so the MbIVa had been designed with a far higher compression ratio. In layman's terms, that meant that the pistons travelled further up inside their cylinders to compress a larger volume of fuel/air mixture and create a much bigger bang. At ground level it produced 300 hp.

It weighed 418 kilos and had a capacity of 23 litres. To put that in context, each one of its six cylinders was twice the size of a whole engine from a modern family saloon. The only problem was that this beast wasn't built for running at ground level. It required fine-tuning to prevent overheating, otherwise six flaming pistons the size of howitzer shells would launch themselves from the bay.

Lou was discussing his project at Brooklands in 1921, and

piqued the curiosity of a quiet, rather precise gentleman wearing a tweed jacket and frameless spectacles. Clive Gallop (pronounced Gallo) rested an elbow in his other hand and gently stroked his clipped moustache as he listened. He rather liked the sound of all this. Before becoming a pilot with the RFC, Gallop had worked at Peugeot on their advanced Grand Prix engines. Latterly at the Air Ministry, he had carried out a performance review of Allied air power. The combustion chamber held very few secrets from him whether at high or low altitude. Lou admired Gallop's reserved intelligence and invited him for a tour of Higham. It lasted about three years.

Plans for Lou's big car were drafted down to the finest points of detail, with certain expedient touches in order to get it off to the races as soon as possible. The sheds were alive with the buzz and flash of acetylene torches, the clump of panel beating and the clink of chains as the colossal engine was finally lowered into its new frame.

To revitalize themselves after long hours in the workshop, Lou and his team tucked into the lavish dinner parties that were a regular feature of Higham life. After a sumptuous banquet prepared by professional chefs, the guests would sally into the gardens, where a string of depth charges went off in the ponds. Fake buildings were also constructed in the grounds. Lou would disappear into a 'building', only for it to be blown to smithereens a few choreographed moments later.

Lou may have inherited his father's racing blood and a tendency to melancholic moments away from the throng, but in all other respects he was the quintessential Englishman – and though boundless in his hospitality, no Englishman suffers an uninvited guest in his castle. During one party there was a thump at the door. Lou, resplendent in full penguin evening attire, went to receive his guest. The man at the door was not on the list. He was, however, a famous racing driver, but one that Lou disliked. Moreover, he was carrying a suitcase with the clear though unstated intention of getting fully loaded and staying the night.

Following the hors d'oeuvres Lou gave Gallop the nod and they removed to a shed. A device was hastily assembled in between fits of hysterical laughter. Gallop carried the device carefully and, while Lou distracted the victim, slid it beneath the intruder's chair and removed the lid of the box. As the guests tucked into the main course and toasted the King's health, the intruder shifted uneasily on his seat. His testicles had tingled at first, but soon his nether regions became distinctly polar.

'Fancy a top-up, old boy?' Lou offered. 'You look a mite pale . . .'

Gallop looked at his timepiece. *Christ*, he thought, *it's been nearly ten minutes . . .*

Unable to bear the drop in temperature any longer, suitcase-man made a dash for the loo, threw down his kecks and gasped at the winter wonderland between his legs. His pubic hairs were frosted icicles. Worse, his cock and balls looked morbidly inert. He flooded the basin with hot water and screamed as he doused them in the scalding liquid. Gallop hid the evidence, and, ever the professional, made a mental note to use a little less liquid nitrogen next time.

He may not have been blessed with subtlety when it came to practical jokes, but Gallop's services as an engineer of genius were ever more widely in demand, so he took on other projects as he saw fit. During the war W. O. had found him to be a kindred spirit in their discussions over the rotary engine for the Sopwith Camel. 'We had talked engines for hours, in the same language and with identical enthusiasm.' Gallop's secret weapon was his invaluable experience at Peugeot's competition department, where he'd been an apprentice to the brilliant Swiss engineer Ernest Henry during the pre-war years. Ernest had configured an engine with twice the number of inlet valves, doubling its breathing capacity. The Peugeot was so efficient that it produced incomparable power for its size. It dominated Grand Prix racing and was paid the ultimate compliment by its competitors: it was exhaustively copied.

In 1919 W. O. had called on Gallop to help him design a new 3-litre power plant for what would become the first purpose-built Bentley. Gallop incorporated Peugeot's inlet design into the engine and complemented it with a second set of spark plugs for each cylinder, resulting in a quicker and cleaner burn. That, along with W. O.'s gift for architecture and alloy construction, enabled them to increase the compression ratio, raising its power yet further.

Harnessed with the excellent chassis designed by Bentley's lead engineer Frederick Burgess, the 3-litre lived up to W. O.'s dream of a 'good, fast car'. After it proved its worth by taking a raft of records round Brooklands in the hands of John Duff, W. O. was persuaded to take on Europe's finest. There was a new event taking place in France in 1923, a marathon-length Grand Prix that would last twenty-four punishing hours round the Circuit de la Sarthe at a town called Le Mans. Bentley was ready for the challenge, but over at Astons there was still work to do before they could even hope to follow in his wheel tracks.

Chitty Bang Bang

After burning a lot of midnight oil, the six-cylinder Higham beast was finally ready by Easter 1921. Lou's rat pack assembled early in the morning and, since this was a team effort, Lou dished out a set of lurid checked caps. They climbed aboard their mounts and set off for Brooklands dressed, according to one observer, like a gang from the 'American underworld'. The gang included the two Cooper brothers, whose experiments with aero engines were nearly as wild as Lou's.

Britain's public roads were still governed by a 20 mph speed limit and Brooklands provided an oasis of speed, with swathes of blooming rhododendrons and other flowering plants setting off the sea of pristine concrete. Large crowds from all manner of society flocked to the Easter Bank Holiday races. Officials with clipboards stalked the paddock to inspect the myriad automotive exotica on display.

Lou's car looked other-worldly. The front half resembled a coffin fit for a giant. A serpentine exhaust, made from an old drain pipe, extended from behind the front wheel to vent into the slip-stream behind Lou's throne. The drive from the monster Maybach engine lurking inside the coffin was taken to the rear wheels by a pair of exposed chain-driven cogs, themselves larger than a regular car's wheels. Although she was shod with the largest cord tyres that Palmers manufactured, the sheer scale of the vehicle made them look like tap washers.

The clipboards demanded a name for the beast. Lou had suggested *Cascara Sagrada*, but someone at the Brooklands Automobile Racing Club twigged that this was an old-school laxative, a cloaked reference to shitting one's pants at speed. Gallop stepped in with an equally oblique suggestion: *Chitty Bang Bang*. The official was no warrior, Gallop figured, and blissfully unaware of the nickname for the paper chit that permitted soldiers to go on leave during the Great War, usually to a place of ill repute.

Lou climbed aboard and readied for the off. He combed his jet-black hair, tied a scarf around his collar and slid on his driving gloves. The boys in their checked caps prepared to stir the kraken. The crank lever used to swing the engine into life was so stiff that it had to be levered using an axle rod. The crowd drew nearer.

Ka-ka-BANG!

Lou plunged his foot onto the throttle, and beneath the long, drab, grey-painted bonnet a sequence of little rockers and arms opened and closed the inlet valves. Fuel splashed into the mighty cylinders and the exhaust gases thundered through the drain pipe to produce an ear-splitting roar. It sounded like an air raid.

Lou eased onto the cement and lined up alongside the other big cars with names like *Creeper* and *Anglo Saxon* for the Short Handicap race. While other drivers tugged nervously on their flimsy linen helmets, Lou took a draw on a cigarette and laughed with Jack 'Shugga' Cooper at his brother's expense as he burned his hand on the searing exhaust pipe. The support crews scuttled away and all eyes fixed on the start flag.

As the flag fell, Lou released the scroll clutch cautiously at first, and then gave it beans. He deftly flicked through the gears and, turning left past the huge Vickers aircraft shed, left the competition trailing in his wake.

Further up the track, a young boy named Ian laid his bike down on the grass and waited for the cars to come round his favourite corner. He was only twelve but the marshals didn't mind him being there. Ian quickly sketched the Members' Bridge crossing from the infield to the top of the steep banking on the outer edge of the track . . . and then he heard it coming. The piercing din of a single engine was strong and pure. The echo reached round the extended curve long before he saw its source. Suddenly, a car shot through the shadow of the bridge, holding a line that was perilously high on the banking by Ian's reckoning.

The concrete at Brooklands had been poured onto a sand base and dried unevenly, creating a series of undulations in the surface. Gallop's main chassis concern for Chitty had been the rigidity necessary for holding the weight of the engine. Shock absorption had taken a back seat. Basically, there wasn't any.

The big car sped across the infamous bumps and took flight. Lou's hair went vertical and the aero engine's decibel level rose greedily by 500 rpm. Watching on, the boy dropped his sketch pad and instinctively shielded his ears from the deafening roar. Lou kept his foot tramped fully home and grinned wildly, making a few mild adjustments with the wheel as he landed and careened away onto the Railway Straight.

The boy answered with a big grin of his own. *My brother's not going to believe this.* Young Ian Fleming picked up his pencil and pad from where they had fallen to the ground and drew a little sketch of the magnificent flying car.

Gallop couldn't believe it either. Lou romped home to a sizeable victory and did so again in the Lightning Round at speeds north of 100 mph. The big bumps had damaged Chitty's fuel pipe, so Lou switched to his Merc to finish off the competition in the 100 Long race. Lou's abundance of natural ability was as clear to Gallop as his absence of fear, and the latter worried him.

Light Car

There were several schools of thought prevailing in competition vehicle design during the early 1920s, and the war's moratorium on motorsport saw an explosion of pent-up energy. Chitty belonged in the Unlimited class at Brooklands, where you could run whatever diabolical invention your heart desired and the officials applied handicaps to maintain competition on the playing field. Grand Prix racing was moving in the direction of smaller, lighter and more nimble machines, leading to a terser form of sprint competition. Vehicle specifications were tightly regulated by the sport's official body, the forerunner of today's FIA, with minimum weight requirements and limits on engine sizes for classes of racing. Meanwhile, over at Le Mans they hedged their bets by allowing up to five different classes of car to contest the same piece of tarmac.

Lionel Martin looked at the shape of the new world and believed more than ever in the concept of the tuned light car as the best value proposition for his fledgling business, and 'decided to start in on the production of a British built fast touring car which would enable patriotic sportsmen to buy British goods . . . a car built up to the high ideals demanded by a small but extremely discriminating class of motorist'.

He dusted off the plans of his original pre-war design at his new premises next door to a milk depot in Kensington, London. He had previously mated an Isotta-Fraschini chassis with a 1400 cc, four-cylinder engine built by Robb of Coventry-Simplex to create the Coal Scuttle. Now, after rigorous road testing, Lionel conceded to himself that his new car handled like a blancmange. The chassis, the vehicle's essential skeleton, had originally been designed by Bugatti for a lighter engine and was flexing too much with the heavier Coventry lump up front. The cure was to make it shorter and add structural support to increase its rigidity. The reboot was designated the AM 270, and it would put the roar into the roaring 20s.

Fortunately for Aston Martin, a woman's touch was on the tiller in the early years: Lionel's wife, Kate, had taken up the role vacated by Bamford. She drew up a stylish monogram for the car's badge. She also introduced an elegant radiator shape with a broadened base and succeeded in renaming the AM 270 the more memorable 'Bunny'. A Devon lass, Kate was a daring driver with an equally heavy foot around the workshop. Parties at the Martin household were endurance affairs, and she could out-drink any man, fixing them with a glassy eye as they slid under the table. Whatever Aston Martin did, they took the party with them.

Lionel and his lead engineer Jack Addis would take to the open road and stir up trouble, daring other drivers to 'tread on the tail of [their] coat'. The resulting tear-ups were useful, if illegal, benchmark tests against rival manufacturers, allowing them to compare handling features and acceleration performance and to probe for weaknesses. A favoured destination was the Brighton Metropole Hotel, where the gang would arrive in a cloud of dust and flit between the bar and the beach.

An enterprising policeman timed Lionel on one of his runs as he prepared to nick him, whereupon Lionel attempted to escape in reverse. For reasons known only to Lionel, the copper, who was on foot, nearly reached the bonnet before his quarry sped off and was subsequently banned from driving . . . again.

With his licence restored, Lionel entered the honed AM 270 Bunny in the Short Handicap race round Brooklands in May 1921 and won it at an average speed of 69.75 mph, marking Aston Martin's first victory at a major event. The Junior Car Club, which governed competitions at Brooklands, then proposed a longer-distance racing format for a much bigger entry of 'light cars', effectively the forerunner of the British Grand Prix, and the new Aston was ripe to be a contender. Gallop was impressed by Bunny's poise in Lionel's hands and felt that the cut and thrust of *voiturette* racing would also be a good fit for Lou, and one that might extend his life expectancy too. He arranged a meeting, and Lou was sold, saying, 'The 200-mile small-car race is undoubtedly the first step towards popularizing motor racing in this country, besides being a practical demonstration of the advance and merit of the present-day light car.' Moreover, he saw in Lionel 'a bloke who would be a real friend'.

Lou signed up to race with Aston Martin and, crucially, he invested in the business. With a nerveless driver behind the wheel, an engine wizard on board and some funding, Lionel could finally raise his endeavours to a more professional level, and he entered a four-car team for the Brooklands 200. Not for the last time would Aston Martin attract a group of peerless diehards to the cause.

For the 200 race Bunny retained the old-style side-valve engine configuration, while Lou and another hotshoe by the name of Kensington Moir drove cars with a newer 'overhead' sixteen-valve engine designed again by Robb of Coventry-Simplex. Unfortunately this proved fragile, and provided barely any increase in power over its predecessor. The Brooklands event was a whitewash, with the French Talbot-Darracqs placing 1, 2 and 3, but it would take more than a drubbing to dampen our team's spirits. Lou wasn't happy with his engine; the bearings that held its pistons were too tight, and he bled speed as they tightened further during the race. But he delighted in the car's agility. He closed out the season by taking the old side-valver for a spin and set records at Brooklands for the standing-start mile and kilometre at 66.82 mph and 60.09 mph.

But Lou's ambition was set higher. He suggested that Aston Martin develop a car to take on the world's premier manufacturers in European Grand Prix racing. Gallop therefore visited his old pals at Peugeot and for the princely sum of £50 commissioned Marcel Grémillon to design a new 1.5-litre engine, albeit one that would utilize the existing bottom end of the Coventry-Simplex in order to save money. While that got under way, Lionel Martin reckoned on settling some scores round Brooklands with ol' Bunny.

In the days before *Top Gear* – in which cars are drag-raced, lapped extensively and subjected to all manner of quasi-scientific road tests – Brooklands provided a test bed for car makers to prove their mettle to the printed media, using both speed over set distances as well as distances achieved within a set time. It was a bit like the Guinness World Records, where the ultimate limit is your imagination, and these records constantly changed hands. Lionel fine-tuned Bunny's engine, assembled his pit crew and picked a team of seasoned drivers in Gallop, Sammy Davis and Kensington Moir for a veritable records heist.

Sammy Davis had been keeping busy as a journalist for *Light Car* magazine, which gave him the opportunity to test all of the contemporary models. Being known as both a safe and a rapid pair of hands, he also raced many of them too. In 1921 he drove for Selwyn Edge, former Bath Roader and now owner of another British marque with a burgeoning reputation, AC Cars. Davis had bagged several Brooklands records for Edge, so Lionel had pulled off quite a coup in bagging someone of his experience. Moir had set a new record up the twisty Brooklands Hill aboard Bunny, so he was another reliable pedaller. And Gallop was Gallop, you had to have him.

Lionel left no stone unturned during his preparation and he did it in secret. He drilled the crew every evening in the build-up to the record attempt. The car was pushed into a mock pit and serviced as though this was the real thing. One man emptied oil from the sump, a second carried in the fresh, while the others unscrewed the multiple wheel nuts, letting them fall where they may and using a fresh set for the new wheel as it was passed in. The repetition led the men to find their own marks and soon the accidental bumping of heads and mistakes were ironed out.

Lionel imposed a one-voice rule, his own, to prevent confusion. Drivers were to follow instructions to the letter, quite a rarity in any era. Blackboards would signpost the number of laps, speeds, control the pace or call them in. It may seem obvious now, but nobody had done this before. It set a precedent that would return at Le Mans in other hands in years to come.

On 24 May 1922 they descended on Brooklands and set out the fuel churns, wheel jacks, spare tyres and all the paraphernalia needed to sustain Bunny for long-distance record-breaking. Lionel marked out the pit area with white paint. According to Sammy, 'the whole car was examined almost with a lens to make sure everything was right'.

Further up the pit lane the rival AC team unloaded their lorry with precisely the same intention.

At 4.25 a.m. Lionel buttoned his overcoat against the chill of the morning and motioned for the first driver to be loaded. The speedway was covered by a thick soup of mist. Few things are more disconcerting for a driver than lack of vision, whether from mist, rain or a steamed-up windscreen. Your imagination plays all sorts of tricks. I shall never forget Jeremy Clarkson screaming into the pits during a foggy twenty-four-hour race at Silverstone to gallantly donate his seat to a doe-eyed James May on the grounds that his own considerable skills should be spared for fairer weather.

Sammy had learned to drive in fog by instinct and to *think* his way round the course using key markers, though in these conditions it still required an act of faith. Not missing a pit signal for a fuel stop or a tyre change were his primary concerns. Brooklands had an abrasive surface that ate through rubber, and a burst tyre would wreck their chances of a record just as effectively as running out of fuel.

Fast pit work was essential for long-distance speed records because every second spent in the pits at 0 mph dragged your average down. Lionel's methodical pit drill was something of a revelation, and a small crowd gathered around the Aston team just to watch them at work. At each stop the tyres were checked or replaced, while the oil sump was drained and refilled. Lionel had fitted a spare oil tank for the drivers to brim the sump manually with a pump if they felt Bunny was running dry. He was a stickler for lubrication, and his engines always had oversized reservoirs and devices for chucking oil where needed.

Under the watchful eyes of sponsors and the media, Lionel's team was flawless – and so were his sneaky tactics. The sun gradually burned through the mist and, as the two teams started trading records, Lionel scheduled his record runs in a way that left AC with no opportunity to respond. The first record to fall was the 7 hours speed average at 76.9 mph . . . followed by the records for 9, 11, 12, 13 and 14 hours – by which time the shock absorbers had lost all pressure and the drivers' view of the track was just a vibrating blur.

As the evening drew in, it seemed that Aston Martin were done for the day. Their tea urn was loaded into the truck and the spent tyres packed away. At 7 p.m. AC pulled their car in for the final time and put it to bed. But Bunny kept running. At 9.20 p.m. she pulled in and received a rapturous reception in the pits, having set a new 19-hour world speed record, covering 1,275 miles. Lionel's team had scooped 22 class records and no fewer than 10 outright world records, leaving the AC team manager to explain to Selwyn Edge how he had been outfoxed.

Grand Prix

Grand Prix racing in the roaring 20s was pure; there were no silly handicap rules whereby officials arbitrarily held back faster cars to balance performance differentials. It was every man for himself, and the greatest drivers in the world drove the fastest cars – Bugattis, Fiats and Sunbeams. For the 1922 season a new set of Grand Prix regulations restricted engine capacity to 2 litres, and that suited Aston Martin nicely. Their new twin-cam 16-valve 1.5-litre engine was ready, and although they still expected to be outgunned on power, it was time for David to take a swing at Goliath.

It may seem laughable now, but in those days most cars only had rear brakes, while Lou's behemoth Chitty had almost none at all. In 1919, however, a little gem of a company called Hispano-Suiza came up with a system whereby a single foot pedal activated brakes on all four wheels. Clive Gallop, always with a nose for the next advance, had worked at Hispano-Suiza for a time and quickly introduced the innovation at Aston Martin. In a giant leap for Lionel's assembled band of mavericks, their Grand Prix cars would be fitted with four-wheel braking. The chassis frame was also tweaked four inches lower on the axles by shortening the front and rear springs, creating a lower centre of gravity for better cornering.

The journey to the Strasbourg Grand Prix was the kind of road trip that Lionel lived for. In contrast to the lavish logistics involved in modern racing, the race and support vehicles were all driven across northern France from the workshop. On arrival, the first hurdle for Astons was that their cars were deemed too light for the minimum weight requirements, but a hefty lunch and a return to the scales soon cured that. The little Astons had already caused quite a stir for having the sheer brass to pitch themselves against Europe's finest *voiturettes*, with a full quarter less engine capacity. And the raucous howl of their exhausts thrilled the crowds, who cheered them on every pass during practice.

The cars moved to the starting grid with Gallop in fifth and Lou back in tenth place based on their qualifying times round the triangular course. The Astons' smaller engines put them at a huge disadvantage on the rather tedious five-kilometre straights, but thankfully it had rained during the night. The routes down the tree-lined boulevards were shining with the residue. It would be slippery work, and that would suit Lou's aggressive style.

Nineteen cars fired up their engines on the start line and the air filled with an oily smog. The grid was cleared and the racers led away by a motorcycle in two-by-two formation for a procession lap, followed by a mass start . . . At least that was the plan. By the time the pack neared the start line most of the drivers were spitting mud out of their teeth and champing to get going, none more so than Lou, and before the motorcyclist had moved aside to initiate the race, the competitors were already buzzing past his ears.

The Italian ace Nazzaro, starting from pole position, showed a clean set of heels to the rest and tore off in his Fiat. Gallop held fifth position. The Aston's new brakes worked superbly into the hairpin, and he was able to fine-tune the attitude of the machine using his foot brake while simultaneously leaning on the hand lever for additional braking at the rear. The Aston's lack of horse-power was less noticeable on the slimy surface, and its light engine helped it through the S-curves.

With the fixed points of a course committed to memory, driving becomes more to do with rhythm than placing the car. The outer limits of performance are then governed by emotion, and if the mood takes you, and the car lets you, your mind elopes into a positive spiral of increasing sensitivity, and it just feels like you couldn't put a foot wrong. Lou was in the mood, fired up and wantonly broad-sliding out of the bends as far as the roadside. His goggles were caked in mud so he removed them and drove *au naturel*, occasionally shielding his eyes behind the bespoke aero screen he had fitted, and miraculously appeared on Gallop's tail.

The two pals were chugging along nicely . . . until their engines packed up. Lou's magneto expired first, followed later by Gallop's. Still, to have stomped around with Europe's finest drivers at speeds of 98 mph was a real achievement. Nazzaro won the race, but was soon to discover that his cousin had been killed when an axle failure flung his car into a tree. It was a sad reminder of the dangers lurking behind the plush curtain of glamour and speed.

The Aston boys investigated the faults behind the magneto failure, with the German engineers at Bosch keen to assert that the installation, not the component, was at fault. While the new engine was more powerful, the fundamental problem with it was that the advanced top half didn't quite match the old bottom. The hand-shake was asymmetrical, and the valve openings had been bodged. When Lou's engine was inspected, it transpired that sixteen of his thirty-two valve springs were knackered.

The next Grand Prix Aston Martin set their sights on was the 200 miles of Brooklands. The Talbot-Darracqs that had dominated the inaugural event the year before were back in force, but Lou and Kensington Moir were equally fancied in the new Aston. Bunny, with *Sunday Times* motoring correspondent George Stead at the wheel, was entered too.

The eighteen cars billowed the traditional haze of smoke on the start line. Seconds before the off, Lou's engine cut out. He scanned the instruments, quickly primed the fuel pump and thumped the starter. The pack set off, its roar drowning the efforts of the little Aston to spark into life, but Lou felt the shudder, pulled away and gave chase. Moir had bolted into the lead, with Henry Segrave and Kenelm Lee Guinness in their once-invincible Talbot-Darracqs desperate to get past him. Moir then extended his lead slightly, while Lou closed on Guinness and, to his disbelief, scythed through into third place.

Jean Chassagne had been favourite to win in the third Talbot until his car stalled on the line and set off with a misfire. He strained every fibre to recover ground until a tyre blowout on the steep banking at the long Byfleet Curve sent him sheering off deep into the wild woodlands of Surrey.

The lead pack rounded the course again and the crowd did a double take. Segrave now led, pursued by Lou and then Bunny. Moir was out, with Guinness in the pits replacing a tyre.

A few laps later one of the valves on Lou's engine cruelly dropped into the cylinder, where it broke the piston, putting him out of contention. Guinness was going again and had powered into the lead, while Stead in the indomitable Bunny had managed to get in front of Segrave and was now circling in a Talbot-Darracq sandwich. With only a few laps to go he was on target for a historic podium finish in one of the most thrilling races ever seen on British shores. But he was running out of gas . . .

Lionel had done the calculations and put in just enough fuel to cover the remaining laps, but he had not reckoned on such a fraught pace. The higher revs were burning more fuel. He conferred with his right-hand man, Jack Addis, and made the painful decision to call Stead in for a refill and concede the position.

'No f—ing way,' came a shrill voice from over his shoulder.

Kate fixed Lionel with her familiar glare and firmly advised him where to put his calculations. The pit crew assembled on the tin roof of the garage to see the race to the end. Stead was driving Bunny for all she was worth, her four pistons beating away to a top speed of 87 mph. Segrave was bearing down on her at over 100. As Stead rounded the longest section of banking at Byfleet for the last time, the fuel drained to the left side of the tank. Bunny took her last sip and, leaving the curve for the final dash to the flag . . . her engine cut out.

Segrave's foot welded his accelerator to the floor as he shot out of the banking at full tilt. Bunny was in his sights and slowing.

The spectators watched on tenterhooks as the little Aston coasted silently along. But at that moment another of Lionel Martin's obsessive traits paid its dividend. His hubs and bearings were always so well polished that they were near frictionless. Bunny free-wheeled gracefully across the line just fifty yards ahead of her pursuer. The crowd went wild, and more than a few eyes welled up in the Aston pit.

But hang on, didn't we forget someone? Chassagne, who had taken a spill over the banking. A rescue party was dispatched. They found a pair of recently vacated canvas racing shoes lying neatly on the track. Their largely undamaged owner was found a little further on, tangled in a sea of thorns along with his mechanic Paul Dutoit.

Aston Martin's cars had proved their pace on the international stage beyond doubt. Lou rounded off the year by battling for the lead of the Penya Rhin GP in Spain with the Talbot driven by Guinness, and placed second. Lou's infectious enthusiasm and his finances had been largely responsible for the rapid development of the cars. However, the twin-cam experience had been painful, and it is telling that Lou's GP machine was refitted with the older-spec engine and put up for sale.

The motoring press couldn't get enough of Astons. *Motor* found their handling 'delightful, docile and suggestive of good breeding, with ample space for two big men and associated luggage'. Built 'for the connoisseur – for the man who likes a machine that is better than those of his associates, so that he may feel a certain definite prestige', their correspondent had not 'encountered any other car which can surpass this performance'. Astons retailed for between £625 and £825 and were deemed expensive, but 'considerably less than that asked today for any other car having a similar range of performance'.

The GP-inspired four-wheel braking system offered 'remarkable stopping power', according to *Light Car*. Their test saw it slow from 60 mph to 0 in just 43 paces, a shorter distance than any other car they had ever tested – and also on a par with a sports car of today. *Autocar* reported that 'it is possible to steer over the proverbial halfpenny with absolute certainty at any speed,' an assertion I find hard to believe but impossible to challenge, while the performance from its 1.5-litre engine made it feel more like a 3. In summary, Aston Martin had produced an unsurpassable sports car.

Chitty's Last Bang

Brave Count Louis had brought Chitty out for another crowd-pleaser at Brooklands and was running full speed ahead during practice. As he sped towards Ian Fleming's favoured curve at the Members' Bridge the front right tyre flew apart, leaving him with no steering. The parapet rail lining the Members' Bridge was the only thing that might prevent the unguided missile from flying off the banking . . .

Lou, with characteristic sangfroid, remembered 'pulling at the wheel unavailingly and suddenly thinking . . . if we hit that parapet, we shall be all right. We did hit it, a deuce of a crash . . . hit it twice, I think. We had bounced off and I thought the affair was all over . . . then we hit the timing box and things were not *quite* so all right.'

'Taffy' Chamberlain was sitting comfortably behind his desk inside the timing box, sheafs of paper neatly stacked and a cup of tea next to the telephone. He heard the bone-jarring crunch from the bridge behind him and was just stirring himself to see 'what's some silly bugger done now?' when Chitty's rear end exploded through the wall.

Lionel Martin rushed to render assistance and Taffy was taken off to hospital, where he lost three fingers but recovered. Lou's mechanic was miraculously unharmed, having briefly joined the flying corps, and a marshal reappeared from his hiding place in a ditch. A mysterious bent shotgun was found, presumably from a failed last stand. Lou had remained inside the car throughout. Lionel approached him and noticed that not a hair was out of place.

Lou simply said, 'Got a smoke, old boy?'

The year 1923 saw Aston Martin enter more events than they sold cars. Lionel improved the twin-cam engine from the GP car and installed it in an ultra-thin chassis in order to break the one-hour speed record of 101.39 mph in a light car. The 'Razor Blade' had a narrowed frontal area with a horizontal shuttered radiator that could be closed manually from the cockpit in order to slice through the air like a bullet. The rear wheelbase was slimmed in order to hide the wheels from the airstream, with a wider front track to provide stability. The widest section of the body measured just over 18 inches. It had a rigid rear axle with no differential and practically no brakes, since these weren't really required.

The bodywork was beautifully streamlined and once a small enough driver was found – cue Sammy Davis – he was sealed inside and pointed towards the circuit. Sammy could see precious little and, as he peered out of the enclosed cockpit, reckoned it resembled a wary oyster.

The Razor Blade was fast and lapped consistently above 100 mph, but it shredded its rubber. With the rear wheels locked into rotating at the same speed through the curves by the crude differential, it tended to push rather than turn, putting additional stress though the front wheels. As a result, Sammy was unable to maintain the high speeds that the machine was capable of without blowing a tyre.

Lou returned to Aston Martin at the end of the year to take another podium in Spain, but when Dr Ferdinand Porsche took over the post of chief engineer at Daimler, he invited Lou to race his new Mercedes. It would fulfil Lou's long-held ambition to emulate his father's exploits at the big car firm and it was common for drivers to compete for different marques during the same season, so he made the fateful decision to take leave of his chums at Astons for a new adventure.

The Merc had a phenomenally high-revving engine for its time, rising to 8000 rpm. It was a 2-litre, supercharged straight-eight-cylinder, and marrying this brute power with a chassis had proved challenging. Racing ace Raymond Mays described its handling as 'appalling' and after watching him wrestle with the car at speed, Segrave remarked that he was 'damn lucky to be alive'.

All this was nothing to Lou, who by then was envisaging a new Chitty with a 27-litre Liberty V12 aero motor . . . and besides, as

Sammy Davis knew from his frequent excursions as Lou's partner in crime, 'it was astonishing what Zborowski could do without real brakes, with loose steering, and with a car that did not hold the road'.

Lou joined Mercedes' crack squad of pro drivers for the Italian Grand Prix and prepared to make history round the high-speed circuit at Monza. The car was barely ready in time for the race and was nearly withdrawn due to clutch problems, but Lou insisted he could work with it. He was dressed impeccably in a black racing suit with LZ on the front pocket and a shirt and tie underneath. While fastening his father's cuff links, Lou remarked, 'I must not copy dear Papa's mistake, must I?'

At the start of the race his car refused to budge until pushed. Lou, undeterred, continued at his usual 110 per cent. He made a stop for tyres and fuel, and, suspecting a lengthy engine start procedure, casually lit a cigarette extracted from his gold case. A touch of serenity pervaded the scene as the mechanics swirled into action. Time to reflect on all other experiences leading to this special moment as he followed in his father's footsteps.

The car was ready. Lou tossed away the cigarette and roared off again, labouring the stubborn clutch for each successive gear change. He settled the vehicle and entered the second curve at Lesmo at over 100 mph. He knew something was wrong as soon as he turned the big wheel. It was ghostly light, as though the wheels were running across velvet . . . Oil on the track. The car gyrated sideways.

Even with Lou's cat-like reactions there was nothing he could do with the steering to prevent the spin. The car veered off the track, collided with a tree . . . and Count Louis Zborowski was killed.

Poor Gallop brought the body of his dear friend home to Higham aboard their ageing car transporter. After turning through the gates of the estate and up the drive, the truck's steering seized up and it could be driven no further.

Lionel Martin called Lou 'a prince of sportsmen', and his death marked a decisive point in the fortunes of Aston Martin. The business was in debt to Lionel to the tune of £31,000 and the artisan approach of producing small numbers of highly detailed cars was clearly in no danger of returning a profit any time soon. Sammy Davis admired Martin for it because 'he absolutely refused to admit anything short of perfection'. Lionel believed that 'the evolution of the sporting car is a thing concerning which no man can say, "It is finished" . . . every month brings some fresh idea for improving the breed.' In addition to the constant tinkering, everything on the cars from their bodywork to the length of the steering column was tailored to suit each customer.

Aston Martin had sold fifty-five such cars to date. Henry Ford sold two million every year. Despite a cash injection by aristocratic investor Lady Charnwood, the company went bust in 1925.

PART THREE

1925–39

The Atom

THE UGLY DUCKLING
AND THE ATOM:
ASTON MARTIN
GETS ITS WINGS

Bertelli

*Lionel Martin was a great friend of mine . . . Actually,
he was a bit of a pompous sort of a bloke. He practically
lived at Brooklands.*

Augustus Bertelli

Augustus 'Bert' Bertelli was born in Genoa in 1890 during
a time of political unrest following the unification of
Italy. His communist father decided to emigrate in 1894,
and the family sailed to England but were blown off course and
landed in Cardiff instead. Bert played rugby, naturally, and with
his powerful frame and rugged good looks could have passed for
an early prototype James Bond. He had a gift for engineering that

he nurtured by taking evening classes on top of an apprenticeship. He followed his star back to Italy and at the tender age of eighteen convinced Fiat to take him into their world-class development team.

In 1908 Bert found himself riding as mechanic to Nazzaro as the great man drove to victory in the Grand Prix at Bologna. The race made a lifelong impression on Bert, as well as another young boy named Enzo Ferrari. Enzo was just ten years old when he watched with his father as an Italian driver won in an Italian car. He vowed then and there to become a racing driver at any cost. Bert returned to Britain to sign up for World War I and luckily for the Germans was turned down on account of a kidney condition.

He spent his time developing engines instead. At about the time Aston Martin went belly up in 1925, Bert was out of a job . . . until he was introduced to a wealthy young engineer called Renwick with whom he started a firm. One of their first hires was a seventeen-year-old prodigy by the name of Claude Hill, a prize-winning draughtsman. Bert took him under his wing and they immediately started work on a new 1.5-litre overhead-cam engine with a view to selling it as a stock motor to other car makers. That's when they heard that Aston Martin was up for sale.

The liquidator forbade Lionel Martin from any involvement in the future concern, and the assets were arranged for disposal. Lady Charnwood had moved the business to some dilapidated aircraft buildings in Feltham, just outside London, and remained determined to save the business. Bertelli took a look around and summed it up as 'little more than a mews garage with a lathe, a milling machine and a power drill'. But after the speed-freak post-war years of Count Louis and his rat pack, the name Aston Martin carried cachet, and Renwick parted with £4000 for a stake in the firm.

The legacy of Lionel, Gallop and Lou was the irresistible allure of Aston Martin, a brand that exemplified obsessive purity in design and an underdog spirit to push the edge of the performance envelope. To any aspiring genius it provided a vision and an energy with which to build their own edifice as much during the 1930s as it does today. It would draw in the best and the brightest, and sometimes, as would be the case with W. O. Bentley, this was whether they intended to be or not . . . Like Lionel before him, foreshadowing everything that would come afterwards for Aston Martin up to the present day, Bert believed that racing was essential to enhancing the product.

Young Claude went at his drawing board with a vengeance, recalling, 'I had a sense of shape, movement, geometry and mathematics, and an intense desire to design and create. I cannot stop.' Everything from the chassis to the gearbox was drawn and built in house to create a completely new car. Bert's brother Enrico created the stylish bodywork that was a feature of this era; his design was arguably the most beautiful in the class.

The engine sired by Bert's team had none of the defects that the marque had endured latterly under Lionel. Although its base was inspired by a proven Enfield pattern, the whole motor was designed in house and featured several advances. The overhead valves entered an inverted V-shaped cylinder head at an angle to create a more powerful vortex inside the combustion chamber.

It would power Aston Martins for the next decade, though Bert drily reflected that it was 'sturdy, but ... far too heavy to give really exceptional performance'. He had a point. At just over 50 bhp there wasn't much whiplash under acceleration. However, it was reliable and therefore ideal for the ultimate test of endurance and speed: the Le Mans 24 Hours.

Quickly establishing itself as a blue-riband event, Le Mans encompassed the thrill of open-road racing but contained it in a closed loop. Driving to the track today you are still struck by its scale, which at eight and a half miles in length is two miles less than in 1923 but still dwarfs contemporary road courses. With max speeds well in excess of 200 mph you spend 85 per cent of the time flat out, and the rest pressing the brake pedal like your life depends on it with so much force that your foot goes numb and you just feel bone meeting metal.

The DNA of road racing involves all creatures great and small competing for the same piece of track, regardless of speed differentials. Le Mans ultimately streamlined the competitors into three categories for small, medium and large cars of increasing horsepower, but it remains a jungle out there – a game of high-speed, high-stakes poker. *Will that guy move across on me, or can I sneak through and gain some precious time?* There's never enough time. Day blurs into night, the sun blinding weary eyes as it drops into the horizon. The twilight is a vague blend of greys instead of the sharp contrast of colours by day. Then night comes, or rather what you can make of it, with the cones of your headlights beaming ahead of you, bouncing like radar off the marker boards, trees and barriers. Did I mention the weather? It has its own ecosystem, and when Zeus pulls the flush it drenches the place, redrawing the map in the minds of the pilots yet again. It's that unrepentant wildness that I loved about Le Mans from the first time I raced there, monsoon and all.

Then as now, the winner is whoever travels the furthest distance in twenty-four hours. If Grand Prix racing was a sprint, this is a marathon.

Bonne Chance

Le Mans had been a French-dominated affair since its inception, with a creeping invasion from Britain. W. O. Bentley sized up the 10.7-mile crooked triangle of road and observed, 'with its long straights, time and again at Le Mans, cars have finished in descending order of engine size'. He set about designing a car with an engine capable of winning such a drag race. Barring some early good fortune, it took Bentley four years to assemble the combination of factors required to dominate the event, not the least of which was luck.

In 1927, the year before the Astons arrived, Sammy Davis drove for Bentley and was larking around at the Hotel Moderne with the other drivers, the famous Bentley Boys. Looking suitably Gallic in his black beret, Sammy peered under his car, *Old Number 7*, and noticed a magpie had taken up residence. The same cursed bird that had shed not a tear for the dying Jesus spelled doom for the superstitious Englanders. But in France magpies were supposed to bring good luck. Which would it be?

Two days later, the Le Mans race was in full swing. Night drew in and feeble yellow lamps shone down into the Bentley pits, where the three cars' lap times were being meticulously logged against fuel calculations. The occasional competitor droned past. The Bentleys had lapped the entire field and if anything the race was a bit dull, until it went quiet.

'Suddenly,' recalled Sammy, 'the watch-keeper stiffened, the machines were overdue, were *all* overdue! He swung round to find W. O. just behind, face white with apprehension. Of all those twenty-two cars not one could be heard, only the noise of a myriad frogs seemed deafening in the still air.'

While the frogs naively chirped away the location of their legs to the peckish home crowd, trouble was brewing on the circuit. The Bentleys had been running in team order, the big bangers with their 4.5-litre engines in front and Sammy in tow with the 3. Approaching the notorious right-left-right curves at White House at 110 mph, Sammy lost sight of the tail lights of his team-mate into the right-hander before lifting off the throttle and gliding through. The honey glow of his headlamps glanced off the big farmhouse ahead to his left. As he prepared to breathe on the brakes for the blind corner, he spotted 'a scatter of earth, a piece or two of splintered wood', the signs of a heavy shunt.

Around the bend lay an unavoidable trap of mangled cars and drivers strewn across the entire width of the track. The previous year Sammy had crashed out of the lead on the twenty-third hour – surely it couldn't be happening again now. *Not again . . . bloody magpie*!

He buried his foot into the brake pedal and swung the wheel. The big Bentley slewed sideways, struck the undercarriage of an overturned rival and came to a stop. Sammy got out and after checking that nobody was gravely hurt, remembered something about a race . . . then took off in a machine that was as bent as a banana and with only one headlamp.

The Cyclops limped back to the pits, where Bentley crew chief Nobby Clark, the former airman who had shared a wartime ditch with W. O., assessed the damage. Broken footboard, bent chassis and front axle but more worryingly a cracked steering ball joint. It may be a well-worn cliché, but 'Winners never quit, and quitters never win.' One look at W. O. affirmed that quitting was not an option. Nobby patched up the car as best he could, reinstalled Sammy and sent it off.

Chassagne now led by several laps in his French Aries, no doubt spurred on by the Brooklands thorns still lurking in his lucky underpants. He seemed unassailable. However, as the race moved towards its climax, W. O. heard a different note emanating from the Aries, imperceptible to everyone else. He pondered it earnestly, then ordered the pit board to be hung over the wall: FASTER.

The wounded Cyclops sped up. The Aries responded, increasing the strain on its engine, which with each lap grew noisier . . . much noisier . . . until eventually it didn't come round at all. Then 'with a roar, "Number 7" swept by, her mudguards flapping in the wind'. Bentley was leading, and the bucket of flailing bolts clung together until the chequered flag.

So began a run of four consecutive overall Le Mans victories for W. O. Bentley. His increasingly large, powerful engines reflected his fundamental belief that there was 'no replacement for displacement'. A defeated Ettore Bugatti called the Bentley 'the world's fastest lorry'.

Despite W. O.'s enduring successes, Bentley Motors was haunted by financial troubles and he would become an employee in his own firm. Nonetheless, his victories placed Le Mans in the minds of the British public as never before and repeating his achievements would obsess British engineers for ever after. Aston Martin's roots were too firmly planted in lightweight agility to take the big engine route, so they continued in the lighter class. Their journey to the top took a little longer and was no less enjoyable for it.

En Avant

Lionel Martin once said that 'during a single competition, a driver may gain more experience than would come his way in a year of ordinary motoring', and as Bert set off for his first Le Mans in 1928, the learning experience began immediately. Astons' 1.5-litre race engines were souped up to 63 bhp and the cars driven over to France. On the route down from Alençon, one hit a pothole and its rear axle collapsed. Bert realized there was a design flaw. He got them to Le Mans and reinforced the axle links. If anything the race track was in worse condition than the roads they arrived on, covered in loose stones and bordered by ditches and majestic forest trees.

During the race Bert kept up a steady average of 60 mph. He closed on a slower car and the driver suddenly pulled across him.

Plus ça change, plus c'est la même chose. Bert swerved onto the grass, clobbered a gulley . . . and ripped the axle off again. The other car ran for eighteen hours until the gear lever snapped off deep inside the transmission and couldn't be repaired. *Nul points* for Aston. But then 'He who never made a mistake never made a discovery.'

Bert went back to the drawing board and beefed up the weaknesses highlighted at Le Mans to produce the International model, designed to strike the right chord at the international competitions the cars were competing in, both as works official entries and in private hands. *Autocar* magazine praised its fierceness and appeal to the 'intelligent' driver. However, its release in 1929 flopped as the Wall Street Crash heralded financial ruin, mass unemployment and dried up the market for sports cars. With slow sales, the company haemorrhaged money and both the Charnwoods and Renwick left the business. Even Claude Hill had to find other work temporarily.

Bert's passion remained undimmed. He pulled in new resources and built three new models to contest the light class at Le Mans in 1931. He increased the compression ratio in the same 1.5-litre engine and squeezed the horsepower up to 70 bhp, capable of 90 mph. The cars were fitted with powerful new Zeiss headlamps that developed minds of their own during the race. After shearing free of their bolts they left the car resembling an inebriated Mr Magoo, looking in every direction but forward up the track. The race officials raised concerns; tensions were high and Bert was beginning to speak Italian, which usually preceded a fit of rage. His solution for offending bodywork was always the same: lashings of rope.

Two of the cars failed to finish, but crucially Bert managed to bring his home in fifth place overall and first in its class. Aston Martin was on life support and this acclaimed result was a shot in the arm. Fittingly, it was at this time that Sammy Davis, an Aston diehard, sharpened his pencil and drew a badge with the words 'Aston Martin' in the centre of a pair of straight-edged wings. His preferred design was black with white etching, while the one adopted for posterity was white on black.

Sales now picked up, but there weren't enough men at the workshop to build the cars. Bert even ended up selling his Le Mans entries to fulfil orders. What was needed, mused Sammy, was a 'thoroughly comprehensive system of manufacture with a full works organization in all the various departments'. At the moment of crisis, investor Sir Arthur Sutherland recognized the quality of its product and bought the company in 1932. The workforce was expanded; sales doubled and Claude Hill returned. Sir Arthur's son Gordon was installed as joint managing director alongside Bert. A lone wolf, Bert would come to find this arrangement tiresome, but the racer in him stayed the course for several years.

Time Machine

It is rare to get an opportunity to climb aboard a time machine but in the course of this project I was granted a special wish. As you enter the showroom of Aston Martin specialists Ecurie Bertelli, who have dedicated their business to the Bertelli era and preserving its history, the first thing that hits you is the sweet smell of vintage oil and real leather. Beyond the shop window lies an array of glittering eye candy that slackens jaws, opens wallets and ends marriages. Garish yellow, red and blue paint schemes assault your eyes with the confidence of the 1920s.

As I made my way round the showroom it struck me that the machines looked delicate, yet perfectly assembled. On closer inspection, they were *infinitely* better assembled than any car I had ever seen. If you look closely at 1990s supercars, as we often do, there are things that make you go 'Mmm' in a bad way: door sills with wonky glass and shambolic seals that you know are going to leak. When you're buying a car it's like a first date and you shy away from mentioning the spinach wedged in her front teeth, but after six months and there's halitosis as well, you want a refund. Not so with these classics. They looked, and frankly were, totally out of my league.

The Bertelli Astons look small, low slung and all very closely related. The same proud frontage that graced Lionel's machines adorns every model, with the top of the radiator slightly narrower than the bottom. Seating arrangements vary between two and four occupants, but around one central theme of primary importance: the driver and front-seat passenger.

The workshop was a wonderful mess. The bare rolling chassis of an Ulster, the racing edition of the mid-30s model, was laid bare to reveal its skinny frame ready for refurbishment. Back in the day a new one cost £750 and maybe £350 for a second-hand one that could now be worth up to £3 million. The ladder chassis, two long beams of metal connected by rungs along the way, looked good to support the weight of a wheelbarrow, but I didn't fancy its chances of holding together through a corner. Ecurie's technical director Andy Bell was cloned from a previous generation of practical engineers. Ruddy-faced, with sleeves rolled up, oily fingers and answers for everything, he laughed at my indignation and doubled down on the car's strengths by showing me a 'drilled' version.

This one had more holes than a Swiss cheese because it had been lightened for racing. As I picked it up and squeezed the slim chassis beams together they bent so easily that I thought they would break. It seemed I knew nothing. Andy pointed out the corner of the frame where the rear wheel attached. Stress fractures had appeared there during racing. The solution was ... to drill more holes, allowing the chassis to flex rather than crack. 'When the wind blows, the grass bends,' and all that.

For eyes accustomed to components manufactured by laser-guided machines, it demanded quite an adjustment to look at a hand-made drive shaft running the length of the chassis to meet a crude worm gear that split power to the wheels. Anything less than millimetre-perfect alignment between the gear faces would result in destruction. That's why one man cut them from solid, assembled them, tested them on the dynamometer, adjusted and then signed them off. Little wonder that Aston Martin built only two cars each week.

Climbing aboard one of these works of art, a Mark II Le Mans from 1934, requires bunching your knees and sliding your legs around the lacquered four-spoke steering wheel. Ecurie's managing director Robert Blakemore slid into the co-pilot's chair, close enough to slap me for errant behaviour. My door closed with a perfect, solid *thunk*. Rob sorted the fuel mixture with a knob on the dash and to my surprise there was an original electric start button. The motor flared into life and pulsed through the straight exhaust like a Harley-Davidson. There was just one problem. The brake pedal was on the right and the accelerator pedal was in the middle. Rob's eyes narrowed in my direction, I released the outboard handbrake and we set off.

It was a cold day, but we didn't put the roof up – you have to admire the British, who buy proportionally more convertible cars than any country in Europe. The ride on the uneven country roads of Buckinghamshire was taut, and the leaf-spring suspension managed the vertical motion beautifully. The motor produced a dignified rattle, but the most noticeable sound was the whine of the gearbox, rising higher in tone with velocity. Changing gear through the tightly arranged gate made a deeply satisfying metallic *snick*.

At higher speed I had to work the wheel continuously to believe in the direction we were taking, like driving a boat. Rob said you get used to it but I still worried about darting across the divide into the path of an oncoming juggernaut. It did explain why it took engineers, even the great Bentley, so long to embrace four-wheel braking; they were worried about the pulling effect the brakes would have on the steering. In the case of Bertelli's machines, that's the reason you can adjust the braking for each individual wheel in order to suit a particular circuit or road conditions, and the ratio of braking front to rear can be adjusted from the cockpit using a hand wheel.

The first obstacle, a roundabout, loomed ahead like an IQ test. The crash gearbox had no synchromesh so it was easy going up the gears but to go down involved double-declutching and remembering which pedal did what. Brake on the right, slow down a bit, clutch in, move gear lever into neutral, blip the accelerator to synchronize the gear speeds to one another, clutch in again and select the lower gear. *I did it*. Then I had to do it again, but this time when I braked I hit the accelerator by mistake and everything went to shit.

My brain went to default settings. I pressed the clutch and put the gear into second . . . The sound of clashing gears announced to everyone on a nearby petrol forecourt that a right tosser was having a go in a vintage car. Rob, a saint in sheep's clothing, kindly offered, 'Don't worry, you can't break it.'

I bet I could.

Modern-day drivers might wisely elect to switch the pedals round to make life easier, but that was not Rob's way because he was authentic. I made a better job of things on the return journey, priding myself that I might almost score a D for good behaviour, until we turned into the trading estate. A truck pulled out in front of us and I stabbed the brakes to pull up. The car stopped on a sixpence and my heart started beating again. As I shook Rob's hand and turned to leave, there was a mutual understanding that he had bestowed quite an honour. I stole a last look at the cheeky little black sports car that begged to be driven, and felt a strong pang of nostalgia for the days when the journey was the destination.

Driving the Mark II Le Mans made sense of contemporary press reviews on what set Aston Martins apart from the herd. Smooth steering, powerful drum-braking and surprising cornering grip from outrageously skinny tyres that enabled journalists to 'amuse ourselves by going into corners at what seemed quite excessive speeds and sweeping round'. Not everyone was blown away, though. Land-speed record holder Malcolm Campbell described his Le Mans as 'slow'.

Bert began looking for ways to increase their power output.

Supercharger

I disliked the easy short cut provided by the supercharger,
which was against all my engineering principles.
W. O. Bentley

Superchargers, or blowers as they were known, are like hairdryers. They boost power by forcing compressed air into the engine's combustion chamber, where it mixes with more fuel to create a bigger bang. The spinning supercharger is harnessed to the speed of the pumping pistons by being attached to their crankshaft – the pedals of the bicycle, to return to the analogy. The spinning parts develop an impulse all of their own and exert considerable force on the engine's internals, as you can experience for yourself by taking a suck on a hairdryer.

W. O. warned that supercharging his engine would 'pervert its design and corrupt its performance', but he wasn't in charge any more. Bentley Boy Woolf Barnato had bailed out Bentley Motors using money from his family's diamond mine, and the board was running the show. At the restless insistence of works racing driver Tim Birkin, a man 'absolutely without fear and with an iron determination', they convinced themselves to fund a development programme. Even Clive Gallop returned as works manager.

A Bentley 4.5-litre was dispatched to Birkin's workshop, and supercharging guru Amherst Villiers went to work on it. W. O. refused to allow Villiers' demonic instrument to be made integral to the engine, so it was fitted externally like a carbuncle. Horsepower shot up from 110 to 175 bhp, and when given a boot-full, the car disappeared faster than a trolley full of loo rolls during a pandemic.

The blower car was entered for Le Mans, and to be eligible to race, Bentley had to manufacture fifty of the model for sale to the public. Due to limited production capacity these came at the expense of making more of W. O.'s Le Mans-winning 6.5-litre monsters. W. O. was scathing: 'The supercharged 4½ never won a race, suffered a never-ending series of mechanical failures, brought the Bentley marque into disrepute.' In conjunction with ongoing financial difficulties, this led to the demise of Bentley Motors and its acquisition by Rolls-Royce. After a short and frustrating spell with them, W. O. moved on to luxury car maker Lagonda.

The blower Bentley did have one important buyer – James Bond, a man predisposed to complicated gadgets and perhaps with enough mechanical nous to keep one running. Ian Fleming wrote the blower into his first three Bond novels, *Casino Royale*, *Live and Let Die* and *Moonraker*. Bond's tastes would mature in time, but let's not get ahead of ourselves.

Over at Aston Martin, under a different owner and striving to get back to the front of the grid, Bert sent one of his Le Mans

Astons down to the Birkin speed shed for some treatment. Inman Hunter was a young apprentice at Aston Martin and watched Bert intently as he brought the doctored machine out onto the forecourt, straightened her up and dropped the hammer. 'To the accompaniment of a whine of straight-cut gears and with a hail of stones like shots from a Bren gun, the car leaped forward, leaving long channels in the gravel surface of the yard.'

A wonderful feeling to be sure . . . until the boys realized that along with a pebble-dashed door to Birkin's workshop, they had also left behind parts of the differential. The offending blower was removed and left to rust 'as a sinister reminder to one and all' of the dangers of imitating Icarus. Bert went with W. O.'s advice instead, and commissioned Claude Hill to stretch his original engine to 2 litres to produce his fastest car yet, the Speed Model.

By this time – the early 1930s – the stock market had levelled out, and a young broker called St John Horsfall, aka Jock, had made a few quid and fulfilled a lifelong ambition by buying a very second-hand Aston Martin International. He completely rebuilt the car himself and won a sprint race at Brooklands, much to the surprise of one of the Aston works drivers competing in a newer car.

Jock had dark penetrating eyes that seemed to go right through you, in no small part because he was chronically short-sighted, and confounded everyone by inserting prescription lenses inside his goggles. He asked the Aston people a lot of difficult questions about geometry and torsion vectors. It didn't take long before they put him in touch with the only person who could answer them, Claude Hill.

Ugly Duckling

Men in suits were a necessary evil to men like Bertelli, who believed them rarely capable of original thought. New owner Gordon Sutherland was different. He was on the ball with experimental technology in America and Europe, particularly the high-end market where US brands like Lincoln were producing slick sedans using the new unibody construction technique. Gordon was also keen to reduce costs by outsourcing the manufacture of some components, rather than making everything in house the way Bert had done. But that and his decision to move funding away from racing didn't go down well with Bert. Having kept the torch burning brightly for over a decade, he resigned from Astons in 1937.

Meanwhile, Claude Hill was slipping into a crater of tedium. His vibrant mind insufficiently challenged, he spent long hours doodling on his drawing board and playing paper basketball. Gordon walked into his office just in time to stop him following a similar course. He wanted to discuss the car of the future. Previously cars had been built on a ladder chassis platform with the bodywork placed on top of it. Unibody made the bodywork an integral part of the chassis structure, thereby increasing rigidity. This opened up a world of opportunity to create independent suspension systems that would have torn the couplings off the bendy bits of yesteryear.

Claude went at the design work like he was inventing a new colour and in 1938 produced a prototype steel chassis with a steel superstructure welded on to it following the unibody concept. It was codenamed EML 132 but soon became known around the Feltham factory as Donald Duck for reasons that were obvious to everyone bar Claude. Despite its high and ungainly roof line, Gordon approved of the design and commissioned a totally new car based on what they had learned with the Duck.

Claude added square-welded cross-sections of steel tube to a conventional chassis structure which enabled it to accept forces from a multitude of directions, and then bonded it with light-weight aluminium body panels. It was ground-breaking work and Aston Martin duly filed the patent. Where yesteryear's chassis had had to be soft enough to twist with the wheels fixed to it, the new structure was so rigid that the wheels could be hung inde-pendently from trailing arms and controlled by coil springs.

The chassis strength enabled Claude to use relatively soft springs for a more comfortable ride and produced more mechanical grip by the wheels' ability to track the road surface. This resulted in higher cornering forces and compliance on slippery surfaces. A subsequent road test captivated *Motor Sport*: 'This is a machine which convinces you within the first half-mile that it is a winner. I have never driven a car which I could handle with greater con-fidence in the wet.'

These intuitive first impressions would go on to be the com-pany's salvation.

Arrangements inside the cockpit came straight from *The Jetsons*. *Why change gear like a caveman with a stick*, thought Claude, *when you could just press a switch?* He employed an advanced Cotal gearbox so that every gear could be selected instantaneously from the dashboard; so if you were heading towards that roundabout where I lost my dignity in fourth gear, you could just arrive, press a button and it would automatically engage first. In effect, two gearboxes operated at the same time with electromagnetic actuators operating selection and a main drive system that dictated whether the car was going forwards or backwards. As a result you had as many gears in reverse as forwards, the perfect getaway motor.

Streamlined aerodynamics gave it the look of an aircraft, albeit a pregnant one. It had a rather gormless expression and retained the comically high roof line of the Duck prototype, but Gordon planned to rectify that in the future.

Fitted with anti-roll bars to control weight transfer and a state-of-the-art limited-slip differential to distribute power evenly between the rear wheels, the new model was exceptionally poised, delivering cornering precision and traction. It was initially powered by the 2-litre engine that Claude had built for Bert's Speed Model, but he was already concocting something more radical to replace it. A formal prototype was produced in 1939 and christened with the more futuristic title it deserved: Atom.

The imminent future, however, would not be devoted to next-generation saloons. A titanic struggle against tyranny was about to envelop every walk of life. It would demand of Britain and the entire world another heavy sacrifice of blood, treasure and technology. But just as in the conflict of 1914–18, the coming war would bring together a new cast of legends in the lore of Aston Martin – as well as include the story of the real 007, whose fictional alter ego would one day power the dreams of a million schoolboys.

One by one, as if by magic, each piece of the puzzle eased into its right place.

PART FOUR
1939–45

DB 2/4 Mark III

SPIES, LIES
AND THE CRUCIBLE OF WAR

Polska

Stubborn, wild, drink-loving, sarcastic, warm, tough, resourceful, inventive and unpredictable. I could be describing the British, but I'm talking about our brothers in Poland. When Napoleon was exiled for the first time he was allowed to take an elite group of his bravest soldiers with him, so he took the Polish lancers.

In 1939 everyone told the Poles not to worry about Hitler, he was just peeping over their fence to have a look at the cows. Poland delayed mobilizing its troops and when Hitler threw two thirds of his forces, comprising 1.5 million men, across the Poles' western border, it necessitated the mother of all fighting retreats. When the Russians invaded from the east with 800,000 men, a lesser people would have popped up the white flag. Warsaw fought to the last, and the Luftwaffe was given licence to destroy the city.

Tadek Marek was building Chevrolets for a factory there at the time. He tried to enlist in the army but was turned down for not owning a second kidney, having left one in Berlin when he crashed during a motorcycle race. Tadek only saw solutions, never problems, and as a gifted engineer the Polish authorities ordered him south to join the resistance on the Romanian border.

He took the company car and drove it very quickly towards Romania, only stopping to pick up his wife along the way – but she had already gone to her parents' in what had become the Russian-occupied east and he couldn't get through. He reached Bucharest, where his car was taken from him for use by the Polish government in exile. Great. Now how would he get her out? Tadek noticed a German diplomat who bore a striking resemblance to himself, and using his finest Berliner accent from his days at Berlin Tech, convinced this man to loan him his car and his diplomatic passport for twenty-four hours. If he was one minute late, he would be reported to the German authorities for theft. Tadek's wife was 900 kilometres away. Even for a renowned rally driver, this was cutting it fine.

Tadek took a route over the Carpathian Mountains where his skills came to the fore, tyres squealing round the relentless twists and turns and brakes boiling from one hairpin bend to the next. He entered the flat plains of what had been Poland just days before and was now Russian territory, bluffing his way through the checkpoints with his German diplomatic credentials. He must have been a good actor because with his rounded features and gentle twinkling eyes he looked more like a young botanist than a Nazi. He found Mrs Marek, and without further ado, they hurried back to Bucharest, making it with one hour left on the clock.

Aviation engineers were being fast-tracked out of the country to assist the Allies but Tadek had to take the slow boat. He volunteered to take a convoy of vehicles to France, little expecting what was to befall that country, and arranged for his wife to travel to Paris via Casablanca. The convoy ran into trouble at the Yugoslavian border, where allegiances were leaning towards the Boche. The guards were girding themselves for something sinister when one of them recognized Tadek. He had watched him slide through that very spot just months earlier on a rally to Monte Carlo, so waved them through.

Tadek made it to Paris and wished he hadn't. It was now 1940 and Herman the German was busy invading there too. When France fell, Tadek knew he needed to get to England, but first he had to save his wife. He cobbled together some travel documents and made it from Marseilles to Casablanca. With no money, he started a business converting gramophone records into jacket buttons. The Vichy French government in Morocco was arresting Polish refugees. The good news was that a week after she arrived, his wife was thrown into the same jail as him. Tadek worked his persuasive powers in French this time, and negotiated their release. From there they went to Tangier, boarded a ship to Gibraltar, then went on to Glasgow and finally reached the Ford factory in Dagenham, where he found work.

Sleepy Britain had woken up to the war, and its factories were toiling to equip the country's defences. A brilliant Polish engineer with total command of German technical terminology stood out from the crowd. Tadek received a tap on the shoulder and was invited to work on some 'special projects'.

Yorkshire

David Brown Senior was a hard lad from Huddersfield, 5 feet 3 inches in height, just as broad and with no discernible neck. He began working as a pattern maker in 1860 when he was seventeen years old. This was an advanced form of joinery that involved carving specialized shapes out of wood, which were used to make moulds, into which molten metal could be poured to produce solid gears and wheels. In a short space of time the business began manufacturing the gears itself.

Working conditions in Victorian industry were severe, with apprentices joining their masters under a minimum bond of eight years. Most employees remained with a company for life and developed an inordinate sense of pride. If David Brown Senior had a religion it was his work, and though he was a taskmaster he was always first in and last to leave. His two sons trailed him round the business to learn it inside and out from day one. One of these boys, Frank, had a son, and in Yorkshire's frugal tradition, named him David. This young maverick would one day redefine the soul of Aston Martin.

Frank would have made a wonderful doorman, with muscles that sprung from his shoulders direct to his ears. He refused to drive, leaving that to his wife Carrie, who was the second lady in Yorkshire to get a driving licence. Frank rarely swore unless he was playing golf. His hands would twist round the club like he was strangling a chicken. He glared at the ball, swung at it and then raged against the laws of nature. Young David seemed to have more in common with Carrie.

David Brown & Sons benefited from a major uptick in business when they adopted helicoidally threaded gears for their worm-drive systems. The worm shaft was like a rotating screw that meshed with and turned the straight-cut teeth of a bigger gear at a right angle. The friction between the old-style gears cost power and the wear was dreadful. By curving the pattern of the screw so that it tended to advance more easily, thrust was improved and wear as well as noise reduced. Demand for the new product soared, and Brown's patented threads found their way into everything from mills to London buses. By World War I they supplied propulsion units for destroyers, submarines and a whole lot more.

David clocked in at the cathedral-like plant every morning at Frank's behest to 'soak him in the business', punching his time card like the rest of the employees. The day began with the opening of mail and Frank's comments on the contents of each letter. Following this marathon, father and son would tour the shop floor. The backs of the workers would stiffen as each station was inspected for irregularities or signs of slackness. On one occasion, Frank found an idler leaning against one of the buildings and fired him. Unable to get a word in, the man accepted his dismissal, and it was only discovered later that he was a contractor from a completely different company.

Frank had tried his hand at building motor cars with his brother and regarded it as a flop. However, Fred Burgess, who had been charged with testing and developing the cars, used to pop young David on an orange box in the passenger seat and take him on test runs across the dramatic hillscapes of the Peak District. This instilled an unshakeable resolve in the boy to enjoy the thrill of driving himself one day. His first step was to build a car.

While working in his father's foundry and ostensibly doing what he was told, David cast a cylinder block for the engine he had designed in his bedroom. He quietly machined the internals and felt that, once he had the engine sorted, 'the rest would not prove too difficult'. Frank caught him red-handed and made David feel like an errant golf ball, but he remained undeterred and, aged seventeen, he built his own trials car based on a Singer which he modified and took racing.

When a special order came into the Park Works from a certain Amherst Villiers for the manufacture of some superchargers, D. B., as we now shall know him, was all ears. Villiers was souping up a Vauxhall TT with twin superchargers, boosting its power to 280 bhp and propelling it to 120 mph. Manhandling such a weapon required seasoned testicles and racing driver Raymond Mays was suitably equipped.

Amherst was looking for a decent place to test the car and D. B. knew precisely the right spot: Holme Moss Road. Climbing sharply from bucolic fields into rugged moorland, the fast sweeping curves reveal the spectacular scenery of 'God's own country' in full Technicolor. The men assembled early in the morning as the sun rose over the peaks. No sign of Raymond Mays.

D. B. offered his driving services, and Villiers thought that really wasn't a good idea, but curiosity always nabs the cat when you stare at a stationary racing car for long enough. Villiers set up a timed stage and D. B. climbed aboard. After a number of consistently fast runs the machine was put away and checked over. Raymond Mays arrived the following day but was unable to eclipse D. B.'s best time.

D. B. immersed himself in the expanding family business and rose to a senior management position aged twenty-two. He spotted weak points in the Coventry-Simplex side-valve engine – *how could we forget them?* – such as oil starvation when it was on hillsides, so he designed a better one with a deeper sump. He had a meeting with Harry Ferguson, who had invented a system for tractors that allowed farm machinery to be mounted on them using hydraulically controlled arms, as opposed to being dragged behind. D. B. at once saw its merit, but it involved wheels, so Frank took some convincing. David Brown & Sons nevertheless part-funded the enterprise that became Ferguson-Brown Tractors.

Reliability issues inherent in the Ferguson design and Harry's tendency to infuriate D. B.'s men by ignoring their suggestions led D. B. to conclude that he should design his own machine in secret. He pulled together a crack team under his draughtsman Albert Kersey, and they worked on a totally new product. Its key parameters would be simplicity, serviceability and style.

The rush to achieving a working prototype is always a fraught one, burning cash on one hand with no sales to show on the other. After another long night failed to produce a runner, D. B. called it a day and sent everyone home. At 2 a.m. he was woken by the dull roar of a diesel engine raging through a straight exhaust. The men had stayed on and finished the job, then driven it for four miles with no lights. D. B. appeared in his suave dressing gown with a whisky in hand to wet the baby's head.

The chassis was built to house modular elements, so that the engine or gearbox could be accessed independently. The engine cylinders contained wet sleeves that could be removed, drastically reducing the costs of a rebuild. D. B. also recruited an engine guru in Alex Taub, who pioneered a more modern, thermodynamically efficient unit that used less fuel. The final touch came from D. B.'s drawing board in the form of a futuristic-looking body, with a proud brow and sleek radiator profile. Function would soon impress, but at launch it was for looks alone that the machine was billed the 'Rolls-Royce of Tractors'. They sold like hot cobs . . . but orders were put on hold when Britain went to war again with Germany.

Gears of War

Men of vision are rarely caught napping. Winston Churchill knew instinctively that the little Austrian with a toothbrush moustache was a catastrophe waiting to happen. David Brown read the white paper in 1935 that called for British rearmament – the armed services were poorly equipped to take on the Germans with their decade head start – and put his house in order. The workforce of 500 would eventually expand to more than 5700 as his factories churned out munitions and parts. D. B. had some 20 mm Hispano-Suiza cannon mounted on the roofs of his factories and drilled his people for air raids and gas attacks. The buildings were camouflaged and decoy lights were deployed to mislead German bombers.

The government created a network of 'shadow factories' that switched civilian engineering firms into manufacturers of the essential war materials. These were dotted around the country in places like Coventry and Birmingham, and might have been harder to find had the British not invited one of the commanders of the Luftwaffe on a pre-war guided tour of the facilities. Strangely, there was no invitation to Huddersfield.

Following the disaster of Dunkirk, Hitler set his sights on the invasion of Britain with Operation Sea Lion. A prerequisite for invasion was to knock out Britain's air defences in order to get his troops across the Channel. The numbers made him brim with confidence. He could fill the skies with 2600 aircraft. Britain could put up just 640 fighters at any one time. The RAF needed to triple in size and get the newly developed Supermarine Spitfire fighter plane into the air as quickly as possible. David Brown & Sons would build their gearboxes. Fighting tanks had been neglected since the Great War. The transmissions in the Churchill and Cromwell models were so brittle that Winston considered parking them 'on the cliffs of Dover and using them as pill boxes'. David Brown's team sorted them out too.

There were other things that Hitler had not considered.

The un-hackable Enigma machine that encoded messages between German commanders had in fact been hacked by a Polish mathematician called Marian Rejewski in 1932. Poland handed the keys to Enigma over to British intelligence, for whom it produced a continuous stream of high-grade information known as

Ultra. Ah, you thought the Brits broke the Enigma code all by themselves. Well, that's Hollywood for you. Ultra provided early warning of proposed actions and painted a picture of the tactics and intentions of the enemy.

Then there was Air Chief Marshal Dowding's system of ground-controlled interceptors. A chain of radar stations tracked enemy aircraft from a range of eighty miles; teams of observers relayed the information to the fighter squadrons via telephone. The system enabled the RAF to allocate its smaller forces for maximum effect and give the Germans a nasty surprise each time they flew over. Then it was down to the few – the pilots, most of whom were British, but there were also a number of wild Poles, Kiwis, Canadians, Czechs, Australians and other Commonwealth heroes to fly what seemed like a never-ending stream of sorties.

The first stage of the Nazi attack was Operation Adlertag – Eagle Day – which aimed to wipe out RAF Fighter Command. Bombers, escorted by fighters, heaved thousands of tons of high explosive ordnance across the Channel and dropped it onto British airfields – or at least the ones they could find.

My dear mum was nearly three years old at the time and living in Sutton, south London. A little blonde cherub with pigtails and ribbons, she was playing in the garden when she heard her mother screaming for her to *run*. She couldn't; she was transfixed by the giant, locust-like monster lumbering low and large above her head. A Heinkel 111 bomber returning to the Fatherland. She recalls seeing the pilot's eyes and leather helmet through the plane's bulbous window canopy, followed by the *thud-thud-thud-thud* of machine gun fire ripping into the ground. Fortunately my grandfather's potted roses were the only casualties.

Goering's bombers yearned to strike a knockout blow, but the bloody British were sneaky. The RAF had noticed that aircraft lined up beautifully on concrete pads were sitting ducks, and began dispersing them into remoter areas away from airfields using ... David Brown's tractors and purpose-built crawlers. The Luftwaffe changed tactics and began targeting the British aircraft industry. This was more worrying. The RAF was replacing its fighter pilots with flyers from Bomber Command and the Fleet Air Arm, but if the factories were taken out, there would be no more planes to fly. Goering had reckoned the Battle of Britain could be won in four days. In the end, maybe he would have to grind the RAF into the ground.

One by one the factories were obliterated. D. B. knew what was at stake. He kept his works humming day and night. His workers, many of them women from the textile industry, worked eighteen-hour shifts. The vital gearbox components that kept the Spitfire in the air ran off the line, polished until they 'shone like silk' by the delicate hands of the women using emery cloths. Those parts harnessed the energy of the Spitfire's mighty Merlin V12 engine, converted its rpm into torque and channelled it into the whirring airscrew.

As the Battle of Britain peaked, D. B.'s factory at Meltham Mills was the last one standing that could supply these parts. The traitor and propagandist Lord Haw-Haw singled it out for destruction in one of his infamous radio transmissions. But the Luftwaffe simply could not find it. Besides being nestled in a woodland valley, its camouflage extended to mock cows on the roof to blend into the countryside. In the end, the closest the Germans got was with a stray bomb that blew up the neighbouring chicken farm, although, many years later, an unexploded incendiary shell was found in the yard.

The biggest air battle in history lasted nearly four months and cost the Germans heavily – 2500 aircrew and 1900 aircraft, to Britain's 544 crew and 1000 planes. For Herr Hitler, who had trampled through Europe, this was his first major defeat, and it put a spring in the step of the Allied counter-offensive. How important were the David Brown factories? A strategic review by the American War Production Board concluded that the Axis powers made 'five fatal mistakes' that cost them the war, including attacking Pearl Harbor and committing America to the fight, opening two fronts by taking on Russia, and Germany's inability to destroy 'a single Yorkshire engineering works turning out virtually all the heavy gears for the British plane industry'.

Lagonda

Another terrifying feature of the German war machine was the Stuka dive-bomber. With the Panzers that swept all before them on the ground, the Stukas formed the tip of the spear of Germany's blitzkrieg lightning strike force. The Stuka's distinctive howl came from a pair of sirens fitted under the wings powered by mini propellers to emit a shrill scream. The intimidating noise caused anti-aircraft gun crews to run for cover instead of fighting back. Dive-bombers were especially effective against ships, as the pilots would enter a steep 70-degree dive and ride their bombs onto the target. The best way to counter them was to fight fire with fire.

A carburettor's job is to mix fuel and air to create a blend with the most efficient burn. It uses pressurized air to draw in liquid fuel, vaporizes it and controls the mixture using a choke to regulate the airflow along with a throttle to increase fuel. Engineers adapt them to suit their engine requirements, and in W. O. Bentley's case, he was used to feeding some thirsty beasts, like his 8-litre. The Lagonda factory where he now worked had switched from producing luxury cars to artillery shells and anti-aircraft weapons, as well as the coolant heads for Spitfire engines. A certain Major General Banks was looking for a way to swat the Stukas out of the sky and at Lagonda he found a way to do it. W. O. and his team created a colossal flamethrower, a carburettor on steroids, that could fire a jet of burning fuel some 250 feet vertically into the sky. Understandably, this device frightened the bejesus out of the Luftwaffe pilots and put them off their aim. The Germans spotted these doomsday weapons being tested aboard ships and set about trying to copy them . . .

Although the war kept W. O. busy, he still sometimes found himself kicking his heels and wanting to do more. Then 'when it was obvious that the tide had turned, I began to think about our post-war car'. The executives at Lagonda were dreaming of selling 'land yachts' again like their lavish V12 drop-head coupé. W. O. didn't expect to find many customers cashing in their ration books to buy those. Instead, he set about conjuring up something smaller and high-performance, powered by a marvellous new 2.5-litre engine. It was exactly the kind of lightweight power plant that Aston Martin had always yearned for.

XX

Ian Fleming had enjoyed a colourful life since he saw Chitty fly past on the banking at Brooklands. Like his hero Zborowski, he quit school at Eton prematurely. Then he went to Sandhurst to become an army officer, but found life there far too restrictive, so that didn't last. He took some exams to join the Foreign Office . . . yawn. Fortunately, in an inspired act born of desperation, his mother dispatched him to Kitzbühel to learn how to speak German and climb mountains. And chase girls . . . drink schnapps . . . and so on . . .

At a finishing school run by former spy Ernan Dennis and his novelist wife Phyllis, Ian learned to spin a yarn at the dinner table and how to write one too. From there he found a career in journalism that sent him all over Europe. Ian's colourful pieces drew in vivid characteristics of local customs, food and phrases. He covered races at Le Mans as well as the political events leading up to World War II.

Ian's older, more perfect brother Peter had enjoyed a meteoric rise from Eton scholar to a first-class degree at Oxford, a successful jaunt in journalism and was cutting a dash as an intelligence officer and pioneer of irregular warfare. He probably paved the way for Ian's introduction to Admiral John Godfrey, the director of Naval

Intelligence, who was in the process of recruiting a collection of first-class minds.

British intelligence had been working on ideas to lure the 'sausage slayers', as one coded message referred to the Germans, into believing a lie that would misdirect their military resources. Godfrey recruited a rogue's gallery of people with the kind of probing minds required to foil enemy plots, create their own, handle double agents, distort the 'truth', gather information and interpret its hidden value or, as he put it, 'push quicksilver through a gorse bush with a long-handled spoon'. He applied a simple filter for selection: 'Only men with first-class brains should touch this stuff . . . it is quite useless, and in fact dangerous to employ people of medium intelligence.'

Godfrey saw in Ian a brilliant wild card that he could throw into his blend of stockbrokers, barristers, classicists and teachers. Although he took Ian on as his assistant, Godfrey allowed him to indulge his creative bent freely and recognized his talent for scheming. The 'Trout Memo' was drawn up in 1939 and contained a raft of devious traps that might appeal to the German psyche as well as keen anglers. These ranged from floating exploding cans of food towards German sailors to distracting U-boats with painted footballs. Fleming included one 'not very nice' idea that he had read about in a novel by Basil Thompson: drop a dead airman behind enemy lines carrying falsified documents about a pending Allied operation.

The memo was circulated among the intelligence chiefs and caught the attention of Charles 'Chumley' Cholmondeley over at MI5. Chumley was an RAF man without wings, owing to his poor eyesight. He was lanky, with neatly combed hair and a moustache that looked like a slapped-on stage prop, all wispy at the sides. His bulky uniform hung awkwardly from his frame like a costume too. Like Fleming, he developed plots, and though some of them were 'so ingenious as to be impossible of implementation or so intricate as to render their efficacy problematical' according to one colleague, they were nonetheless potent.

Chumley was also secretary of the XX (Twenty) Committee, also known as the double-cross committee, which was charged with counter-espionage projects. Chumley became aware that the Germans were susceptible to believing fake dispatches after an incident on the coast of Spain. The body of a dead British officer had been found by Spanish fishermen and turned over to German intelligence, who disregarded the invasion plans they found in his pockets. When the Allies did invade North Africa during Operation Torch, just as the dispatches had foretold, the Germans kicked themselves for missing the warning.

Thanks to Ultra, Chumley knew the Germans were ripe for the plucking. He pitched his own idea of depositing a body in the sea, where the Germans would find it along with fake top-secret documents. Chumley was given the green light to make a feasibility study and, since it involved the sea, liaised with the Royal Navy representative at XX, Ewen Montagu. 'Monty' was an eagle-eyed barrister who could dismantle an opponent's argument in court as if he had written it for them. He analysed the plan intently, pulled it apart and put it back together again.

Monty had lost five of his teeth in a riding accident that left him with a crooked smile and a convenient hook for his pipe. As he contemplated his German counterparts swallowing this rotten bait, a diabolical smirk grew across his face.

Operation Mincemeat

The year 1943 marked a turning point in the war, with the Allies' success in North Africa and Hitler's failure on the Eastern Front. The next step for the British and Americans was to cross the Mediterranean and reclaim 'fortress Europe' from the Nazis. This required a major assault, and it was painfully clear how ferociously the Germans would defend themselves. It was equally obvious where the assault would go in – as Churchill pointed out, 'Anybody but a damn fool would know it's Sicily.' Hitler thought it might be Greece.

Sicily is the closest part of Europe to North Africa and, sitting at the toe of Italy, was the gateway to the continent and the Med. The assault would require the biggest Allied invasion force ever assembled and the Germans would surely spot the preparations, but if they could be misdirected into defending the wrong target, it could be the difference between winning and losing the battle.

The plan to fool them was code-named Operation Mincemeat in a macabre reference to its leading actor: fictional deceased Major Martin of the Royal Marines. Ostensibly the major had been travelling to North Africa by plane when it crashed into the sea. The major would wash ashore in Spain, where Nazi sympathizers would hand his briefcase over to the Germans. The contents would make for fascinating reading. In addition to his outstanding debts with Lloyds Bank, recent engagement to be married and penchant for West End theatre productions, the major was privy to secret information about the proposed invasion that he obliquely spilled in his correspondence with X and Y. *Mein Gott, zee Engleesh vont to attack Greece!*

In reality the major would be transported up to Scotland, where he would be secretly loaded aboard a submarine. From there he would be taken to the waters off the coast of Spain and floated towards land on the tide.

With the plan agreed the next step was to find a willing corpse. There was no major, not yet, and Monty knew all too well that body-snatching was completely illegal. After some brainstorming with Chumley in the smoke-filled rooms of the Admiralty, it dawned on the boys that finding the right body would be harder than it sounded. The cadaver couldn't have been shot because that didn't fit the story, and not many families would be willing to relinquish the remains of their dearly departed without a good reason, which was impossible to give.

Monty turned to an old colleague who was a coroner and remarkably for this story went by the first name of Bentley. He was an obliging patriot, having given the cause of death as 'heart attack' for a captured Abwehr spy who killed himself by shoving his head in an oven. Monty stressed the national importance of finding a fresh body. Bentley duly found a suitable candidate in the sad case of a Welshman named Glyndwr Michael who had ingested rat poison while living rough on the streets of London. He was placed in cold storage and the operation stepped up a gear.

Major Martin, as he was thenceforth known, had to look the

part and needed credentials. With no photo booths available, Monty had a go with his camera but lamented, 'It is impossible to describe how utterly and hopelessly dead any photograph of the body looked.' Things were slipping into the farcical, but by good fortune Monty encountered a man over dinner who bore a striking resemblance to his icy friend at the morgue and smooth-talked him into having his picture taken.

While Monty and Chumley were undoubtedly the brains of the operation, they needed a man of action to help them bring it to life. Enter the third man, Jock Horsfall, who we last met speeding round Brooklands before the war.

Jock was as comfortable riding a motorbike sideways round a grass racetrack as at the wheel of an Aston Martin. His mother had been a driver for MI5 chief Sir Eric Holt Wilson during World War I, and Jock had followed the beat of the same drum. He was assigned to A Division of MI5 and fulfilled various roles. He spied on sensitive British installations to test their security, while on the flip side of the coin he fed fake news on military hardware to the Nazis through porous networks in Spain. As a racing driver his skills came in handy for hot-footing agents on secret missions.

Jock popped round to the morgue and joined the group dressing the body from underwear up. The poor major looked understandably pallid and was frozen stiff. They managed to dress him until it came to footwear. His feet were not playing ball, so they had to be thawed with an electric fire in order to fit the boots.

Major Martin was given a complete back story in the form of snippets of evidence from his everyday life. His pockets contained meticulously sourced receipts in case the Germans followed up to verify their authenticity, from the bill for an engagement ring to those for his uniform. The latter included a cash receipt from Gieves of London, where officers enjoyed the privilege of running a tab, and Monty only realized when it was too late that this might alert a German snoop.

In addition to love letters and family correspondence, he carried written orders specifying that his services were required in preparation for a future operation, based on his expertise in amphibious assault by landing craft. As to why he was carrying a secret letter between two high-ranking commanders, this was explained away as being to prevent interception of the information's transmission by wire.

With his pockets and briefcase brimmed with bogus info, it was time to get moving. A special steel canister was designed by Charles Fraser-Smith at Q Branch to transport the major in such a way as to be inconspicuous, but adequate to protect the contents and keep them cool. One end was sealed while the other had an air-tight hatch. It was a suspicious-looking item that required third-party manufacturing, so Jock ordered twenty-four to mask its true purpose. It was packed with dry ice to expel oxygen and prevent decomposition, then the major was loaded.

This seems like a good opportunity to step into Q's lab. Mind you, looking back on the ingenuity gracing the history of Aston Martin, perhaps Q has really been riding with us all along, turning his hand to everything from automotive to aero engines, flame-throwers and nut-numbing party tricks.

On paper Charles Fraser-Smith was a civil servant part-timing for the Ministry of Supply's Clothing and Textile Department; in reality this was a screen for his deadlier brand of fashion for Section XV of the SOE. His Q gadgets took their name from the merchant ships during World War I fitted with concealed weapons in order to lure U-boats to the surface and then sink them. Similarly, agents and commandos had to sneak weapons as well as information and signals equipment in and out of enemy territory. Charles's speciality was converting everyday objects to perform that task.

He had attended Brighton College, where he was deemed 'scholastically useless except for woodwork and science and making things'. Teachers eh? He had a gift for grasping detail as well as the mindset of his opponent. His secret document container opened using a left-handed thread that supposedly made it invisible to the 'unswerving logic of the German mind', which would never consider turning the screw the 'wrong way'. Focusing mostly on small items in his little office at St James's, he hid mini-cameras inside cigarette lighters, created slip-ons to go over army boots that left footprints of bare feet, made a pen that turned into a gun and crafted shoelaces out of steel to be used as garrottes. His department even produced exploding rats, so perhaps Roger Moore wasn't the first spy to emerge from the jaws of a crocodile submersible like the one in *Octopussy*.

Jock picked everyone up in his square Fordson van, the one he used to transport his racing car, and they set off, looking for all the world like Postman Pat's stag do. On their way through London the boys spotted a queue of cinema-goers lining up for a spy flick.

They fancied the contents in the back of the van would make a far more dramatic story, and pictured the faces of the civilians if they were treated to a glimpse of the corpse inside their canister. It had been a tense few weeks of planning and preparation up until this point, and once one started laughing the domino effect was unstoppable. Jock nearly crashed into an oncoming tram.

Monty and Jock took turns at the wheel, driving in stealth mode with the headlamps dimmed, and at one point they vaulted over a roundabout that suddenly appeared in front of them. They kept the windows open to listen out for enemy aircraft but thankfully the engine tones were of the friendly variety, and they arrived safely ahead of schedule at Greenock port just outside Glasgow.

After a spot of trouble lowering the 400 lb canister of 'meteorological instruments' into the launch ferry, Major Martin was received aboard the submarine HMS *Seraph* and departed on his mission the next day. Lieutenant Commander Jewell brought the submarine to the surface just off the Spanish coast at low water in the early morning. After sending the men below, he revealed the plan to his senior officers and read a brief sermon before releasing the body into the water.

The poisoned bait was afloat. To make it a little more appetizing, Monty had included a woeful account of the failed Allied raid on Dieppe which he felt the Germans would find too irresistible not to share with their superiors. As for the invasion plans, Monty used the genuine operational name Husky but switched the real target to Greece. He also referenced a bogus operation, Brimstone, that indicated Sardinia. The real sweetener, from an Allied perspective, was the pretence that Sicily would be portrayed as a decoy target,

BEN COLLINS

a perfect double bluff, so that indications of the genuine invasion would be ignored by the Germans . . . But were they?

Major Martin's briefcase was returned to London by the Spanish, its contents 'unopened'. The letters, however, were surprisingly dry, with evidence of having been copied. Ultra began to whisper. Hitler ordered seven divisions to defend Greece, including the elite I Panzer Division under the command of General Rommel, ten further divisions were sent to the Balkans and troop numbers in Sardinia were doubled. Nothing was sent to Sicily. The British chiefs of staff could hardly contain their excitement and wired a message to Winston Churchill in Washington that read: 'Mincemeat swallowed whole.'

British and American forces invaded Sicily on 9 July 1943 and conquered the island in thirty-eight days, a third of the time forecast and with far fewer casualties. Even after the invasion the Germans clung to the notion that Major Martin's letters were 'above suspicion' and that the Sicily campaign was just a ruse. A significant portion of Germany's forces were taken out of the war for a time and thousands of lives were saved by our quirky intelligence team. Ewen Montagu was overwhelmed by a sense of relief that he found 'impossible to describe'.

Commando Reloaded

For Ian Fleming and Godfrey at the Admiralty, it must have been satisfying to see their 'Trout Memo' lead to such a successful operation, although not everyone was as grateful as they might have been. Montague vaguely recalled to Godfrey, 'I honestly don't remember your passing on this suggestion to me,' but, 'what you said may have been in my subconscious and may have formed the link.'

This was an occupational hazard for an ideas man like Fleming. But he was a man of action too, and like the character he would write into existence in 007, he railed against the limitations of an office-bound desk. Thanks to Winston Churchill's declaration to 'set Europe ablaze' using unconventional warfare, the creation of elite commando units was much in vogue, and most of the military departments created their own groups of hooligans to meet the call.

Until there was a sufficient fighting force to take on the mighty German war machine, Churchill was keen to maintain an effective resistance to German occupation across Europe and saw these irregular warriors as the means to encourage it. The aims of their activities were to sabotage key enemy installations, disrupt communications, gather intelligence and divert German military assets onto internal security. Initially it was largely about blowing stuff up. The ulterior motives were to raise Allied morale and annoy Hitler into making mistakes. His annoyance, perhaps even fear, was confirmed soon after the first raids when he ordered that any captured commandos would be put to death.

Surprisingly, there was no shortage of volunteers for this duty and the recruitment process for commandos was both selective and vigorous. Applicants came from the cream of the military as well as civilians and refugees fleeing occupation or looking for revenge. The Special Operations Executive put their recruits through a four-stage meat grinder to generate the right stuff. Passing an interview led to physical assessments in the highlands of Scotland, with blister-bursting outdoor excursions and long, cold nights learning to survive while observing enemy positions.

Formal parachute training took place at RAF Ringway, although when this was unavailing the SAS took to throwing their men out of trucks at 30 mph. The impact served two purposes since it removed through injury those who might prove to be unlucky, while honing the skills of the rest. The finishing school at Beaulieu taught any remaining agents their clandestine trade for planting and improvising explosives, forgery, breaking and entering, assassination and abduction techniques.

Instruction in unarmed combat was provided by Fairbairn and Sykes, former Shanghai cops and creators of the commando dagger. Sykes' classes in throws, joint locks and the infamous Judo chop always concluded with the same *aide-mémoire*: '*and then kick him in the balls*'.

The end product was a resilient individual, independent of thought and ruthless in execution. A high-stakes poker player by contrast to the calculated, chess-playing codebreakers at Bletchley Park. It instilled a pithy, if macabre, sense of humour that could be detected in mission reports such as one from the legendary Anders Larssen (VC), who reported one SOE mission to his superiors thus: 'Landed. Eliminated Germans. F***ed off.'

Fleming recruited his own branch of deadly commandos for Naval Intelligence in 30 Assault Unit, his 'red Indians', for the purpose of stealing sensitive information such as enemy codes and ciphers. Their chief knew too much to be put at risk alongside them on operations and Godfrey forbade it. What is certainly true is that Fleming warmed to these big personalities. At the end of the war when he decided to write 'the spy story to end all spy stories' and developed the character of James Bond, 'he was a compound of all the secret agents and commando types I met during the war'.

And ultimately only one car on earth would be good enough to match his hard-hitting style.

PART FIVE
1945–59

DBR1

BENTLEY, BROWN AND BOND: SHOOTING FOR LE MANS

David Brown

The war finally ended and dread of the air raid sirens faded. Even though Britain's economy had been shot below the waterline and would never be the same again, her factories resounded with pride and people were allowed to dream again. Love was in the air too, followed by a baby boom. For an entrepreneur like David Brown, here was an opportunity to create a product that delivered romance and pure escapism.

At Lagonda, W. O. was keen to bring his new 2.5-litre creation to market but he was stumped yet again. The newly installed Labour government imposed strict limits on raw materials that were too restrictive to make tooling up a new automobile worth the investment. Lagonda was facing bankruptcy.

Over at Aston Martin, the factory was returned to its owners by the Ministry of Aircraft Production. During the war years Claude

Hill had designed the tooling used to manufacture joysticks for the Spitfire, a coincidence as my grandfather, Ernest Collins, built the tooling for their undercarriages. As an example of the computational brainpower these men wielded using only slide rules compared to regular mortals like myself, I recall Grandad being collared by the police.

He was guilty of a number of infractions involving other road users that he had not seen coming that had left his Talbot Sunbeam looking more pockmarked than an asteroid. Ernest was as blind as a bat but when a policeman walked him into a car park, he saw what was coming. He memorized the registration numbers of every car they passed until the copper asked him to turn around for an impromptu eye test. 'Mr Collins, please tell us the registration number of that red car.' Using the colours to guide him, Ernest trotted off the plate numbers of every car he pointed to and was released back into the wild. Lord of the slide rule, he could multiply ten-digit figures instantly in his head and so memorizing fifty licence plates was a walk in the park. Sadly I failed to inherit these attributes.

During wartime Gordon Sutherland had put 90,000 miles on the Atom, logging valuable testing notes that he shared with Claude Hill, who, with his photographic memory for dimensions, jumped straight back in with front suspension revisions. Claude also completed work on his new 2-litre engine with an elegant system of push rods controlling the valves of its four cylinders. Mounted low inside the Atom, it was capable of running at up to 90 mph at a respectable 22 miles to the gallon. But Astons faced the same problems as Lagonda; after two years, Gordon was forced to sell.

David Brown, on the other hand, was above such setbacks. When his was the only factory producing vital gear parts for the Spitfire, he was told he would have 'all the steel he needed'. Perhaps helping to win the Battle of Britain, putting the army's tanks into gear, mobilizing the RAF with tractors, building the bombs they used to take out V-rocket facilities and (rumoured) work on the Dambusters' bouncing bombs had warmed the hearts of the grey-suited men in Whitehall enough to earn a few post-war privileges.

D. B. was still building tractors, but looking to expand his business. He was leafing through a copy of *The Times* when he spotted a car company for sale. Investigations revealed it to be Aston Martin. D. B. took a trip to the factory to look at not very much, but Gordon had one ace up his sleeve and handed him the keys to the Atom.

D. B. would have immediately appreciated the sophistication of the Atom. The smooth silver lines were striking, and he could forgive the unusually proud roofline, this car being a one-off demonstrator. The red leather seats were comfortable but assuring, the cabin space surprisingly open with extraordinary breadth of vision through the split windscreen. At cruising speed the semi-automatic gearbox worked easily and there was not a trace of rattle from the bodywork. For the real test he headed north towards the Pennines and a run across Holme Moss, one of Britain's most iconic climbs.

The cushioned ride might have meant that the Atom would handle like a bus, but D. B. could feel that wasn't the case long before he reached Yorkshire. In the long sweeping bends up and around the valleys, the Atom clung on far beyond the point where he expected the front end to break away. There was some roll, yes, but only enough to tip the nose towards a corner and feel it take an assured set. Changing direction caused a little more commotion, but nothing you couldn't dial out with a roll bar. It was a bloody good car, if a 'bit underpowered and not very good-looking'.

He bought the car and the company for £20,000. A couple of months later he was tipped off by a distributor that Lagonda was going into liquidation and would be put up for sale. D. B. did a tour of the factory in Staines, saw some pretty cars and W. O.'s new six-cylinder engine. The extra cubic capacity would be ideal for the new Aston, though D.B. presumed that he wouldn't be able to afford to buy such a large enterprise as Lagonda and was about to dismiss the idea. But by good fortune D. B. knew the receiver, who accidentally on purpose left the sealed bids on the table while he left the room to tie his shoelaces. D. B. just pipped the best offer and Lagonda became his for £52,500. The combined talents within these two firms would propel Aston Martin to dizzying new heights.

DB1

D. B. envisaged the new Aston Martin as an open-top four-seater sports car, so Claude set about slicing the roof off the Atom and made up for the loss of rigidity by strengthening the chassis. News of Astons' resurgence travelled fast. Jock Horsfall returned to the scene and was overjoyed to be hired as Claude's test driver to develop the new car.

To say that Jock didn't do things by halves would be an understatement. In between spying missions during the war he was always busy. During the Blitz he had manned the roof of MI5's building at Wormwood Scrubs and *kicked* off the incendiary bombs raining down on them; one eventually got the better of him and he fell through a skylight and broke his neck.

Jock had privately entered, and won, the Formula B category at the Spa-Francorchamps Belgian Grand Prix in 1946 with his old and heavily modified Aston Martin Speed Model. Journalist Dennis May had watched him 'put the job together like a watchmaker and then drive it as though his one desire was to smash it irreparably'. As soon as Jock tested the DB1 prototype in 1947, he knew it was a winner. He would take it onto the roads around Suffolk, mark out two distant points and time his runs between them for days on end.

Jock paid particular attention to the steering arms, which he carved from solid steel in order to reduce flex, providing pinpoint accuracy. His hands read the road like Braille through the ropebound steering wheel. He discarded the standard seat and built his own closer-fitting design in order to keep his body tightly fitted to the car, allowing his body to become one with the chassis and respond instinctively to the animated cornering forces. If one bend highlighted a malady in the handling, Jock would feel it like the princess with the pea. Repeating the bend for hours, making subtle adjustments to the suspension in search of the sweet spot, Jock took the car ever closer to its performance limits. Testing is a curiously satisfying process, chipping away the dead wood lap after lap, closing in on tiny details to unlock a car's ultimate potential.

Eventually this state of nirvana was precisely where Jock found himself. Everything from the tyre pressures to the spring and damper settings was working in unison. The braking balance was perfect, and he could adjust the car's cornering attitude by simply bleeding a pound of foot pressure from a tyre. When you have that level of connection, the car is screaming at you to race it.

Big motorsports events were few and far between immediately after the war, but in 1948 a major twenty-four-hour race at Spa was being organized. Jock just had to convince the new boss. D. B. saw potential in the bodyless prototype racer, but there were only weeks until the race in which to build a competition-grade car, and his gut feeling was that the four-cylinder lump was underpowered. But he went for it. All hands were put to the mill to turn out a DB1 racer. It was rushed over to Jock's workshop for final build, fine-tuning and set-up. D. B. sent an apprentice to assist Jock, but when the lad returned to the factory he explained that although he had been allowed to touch the car, it was only for cleaning purposes, Jock's mantra being 'if a job needs doing right, do it yourself'.

Spa

The Spa-Francorchamps circuit carves up, around and back down a steep tree-filled valley to give it a unique roller-coaster quality. Nestled deep within the Ardennes forest, it is one of the oldest venues on the F1 calendar and regularly voted the drivers' favourite. In Jock's day, the circuit was longer and faster, but it had the same village feel. Commitment was vital for throwing speed into the long, often connected sweepers. Accuracy counted because invisible contours in the road rubbed the car off line and cost time on the stopwatch.

The teams operated from an inner paddock that sloped from the tight La Source hairpin down a toboggan chute into Eau Rouge, where a tree-lined bank loomed close to the outside of the track. In full view of the pit crews, the cars gracefully turned left at full tilt before the course bottomed out, then veered uphill to the right, compressing every sinew, and finally broke over a sharp crest onto the straight. You could tell who had balls just by listening to the exhausts, and some drivers stood out.

Twenty-four hours is a long time to stay awake, harder still to remain alert. Mental fatigue creeps up on you. One moment you're in the zone, scanning for brake markers, the next a fleeting thought about something as random as Christmas shopping breezes through your subconscious, and if you don't get a grip within a matter of seconds it can all end badly. Jock was used to sleep deprivation from his spy days and he shared the driving with an accomplished racer called Leslie Johnson. Pitted against a field of thirty-nine, the little Aston belonged to one of the smaller classes and was not the fastest, but it was built to last.

A heavy downpour coated the track with rain before the race began, a familiar occurrence at Spa, which has its own climate and can be bone dry at one corner but soaked at the next. At the start a Delage driven by Bob Gerard had the lead until a factory-entered Ferrari took over. The night brought more rain and plenty of casualties on the slick, at times flooded, course.

Gerard spun at a fast section, 'shut my eyes and found myself among the pine trees'.

The Aston Martin with its soft, compliant suspension revelled in the conditions, and Jock and Leslie picked off their opponents one by one. The only real threat to victory seemed to be coming from the pit next door. A privately entered Aston Martin driven and owned by Dudley Folland was just half a lap behind and trading fast times with D. B.'s men. It was a small but professional outfit under the management of a tall character called John Wyer who had the resigned expression of a mortician.

During the night D. B. approached him and suggested a deal to guarantee the finishing order, to which Wyer cryptically replied, 'Does that mean that *we* win and *you* are second?' This was not what D. B. had in mind. Wyer discussed the matter with Dudley and they agreed to play nice . . . for now. A short time later Wyer's car came in and he borrowed a high-voltage lamp from Astons' crew chief Jack Sopp. The rain had soaked the connections, and when Wyer pressed the switch he got more than he bargained for and danced an electrified Macarena across the pit lane.

At midday on Sunday the tank on Dudley's car split going up Eau Rouge, spraying slick fuel onto his rear tyres and spinning him off into the weeds, leaving Jock and Leslie clear to run to the flag. When the clock struck 4 p.m. on Sunday they crossed the line to win not just their class, *but the entire event*. It was a strong vindication of Claude Hill's design and Jock's perfectionism that took David Brown completely by surprise. Claude couldn't believe it either, and after the race 'got totally drunk'.

The whole point of motor racing for factory-backed works

teams like Aston Martin was to sell their latest cars. New DB1s were quickly assembled to be put up for sale, and even Jock's winning racer was converted into a road car to be flogged. The DB1 was styled by ex-Lagonda designer Frank Feeley, who incorporated ideas from his pre-war work on the Lagonda V12. He used elegant proportions and longer body lines than the Atom, along with a less surprised expression. It launched at the London Motor Show for £63,000 in today's money, which was strong for a nation of people with 'ration books, rickets and unemployment'. They eventually managed to sell fifteen.

D. B. had bought Lagonda for its talent and was keen to use it. Claude's engine, while brilliant, needed investment and development in order to grow it from a four- to a six-cylinder power plant, so D. B. decided instead to plug W. O.'s six-cylinder unit straight into the new DB2 and ended up having a 'frightful row with Claude about it'. As a result we bid farewell to the dynamic duo of Jock and Claude, who both felt it was time to leave the firm, but not without a final flurry.

Jock had one last score to settle – with himself. He entered his own Aston in the following year's Spa 24 but this time planned to drive the entire race himself. Paul Frère agreed to stand by and take over the driving if Jock collapsed. The car had enough fuel to run for three hours without stopping, which meant that Jock would have to drive eight long stints during the twenty-four hours. The longest I've ever driven continuously in a race was four hours, and that was knowing there would be an equivalent break afterwards.

Jock came in for his first stop. He felt perfectly normal until he went to climb out of the car and his body simply wouldn't budge.

His muscles were locked so stiff with lactic acid that he had to be helped out before he could hobble around to ease his limbs. They dipped the fuel tank and realized that since they were using less fuel, his stints would last five hours non-stop from then on . . . Jock would lose strength inexorably for the rest of the race.

He managed to continue on a diet of chocolate and boiled eggs, a dodgy egg producing a nasty turn after dark. By 4 a.m. he had moved on to the hard stuff with tonic water and brandy. The cold of the night and the relentless pounding over kerbs, the shifting of gears, the squeezing of brakes, the saddle sores and the darkness were taking their toll on him. Dizzy, dehydrated, cold, depressed, Jock began ignoring pit signals and was nearing the point of no return. At that point your actions in the car become robotic and your sensory perception borders on hallucination. It looked like Frère might have to take over, until the sun popped its smile over the horizon and Jock warmed up. At the next pit stop he popped a Benzedrine tablet that he had been saving and steeled himself for the final push.

The Bennie helped Jock refocus, but his muscles were so jacked with lactic acid that he struggled to turn the wheel all the way over for the tight hairpin, regularly running wide onto the grass. But eventually he slogged his way to a fourth-place finish with a faster time than his race-winning performance from the previous year. The crowd gave him a standing ovation.

Jock recovered from his ordeal with a hot bath and a bottle of brandy, although another competitor stole his prize money, leaving him with nothing to pay his way home. When you had the Gestapo on your tail, the last thing you wanted was a blown motor; Jock had been the go-to guy if a resistance fighter needed a reliable souped-up special, and who knows how many lives he had saved with his unique attention to detail. So it was no surprise that some loyal 'friends from the continent' helped to pay his way home. Two months later Jock undertook his final mission. Driving an evil-handling, supercharged machine made by ERA (English Racing Automobiles), he was killed at Silverstone.

Rise

Aston Martin thrives on big personalities, although inevitably the energy of youth mellows into experience, which eventually hardens the attitude. The Lagonda men coming across to Aston Martin had been through a lot of change and none more so than W. O. Bentley. Having lost control of the company bearing his name, he had been brow-beaten by men in suits at Rolls-Royce and had hoped for salvation in Lagonda, which promptly went bust. Bentley stuck around at Astons for a short while, but the prospect of doing it all over again didn't appeal and he moved on.

The DB2 was based on the chassis from the DB1 but with a shorter wheelbase, while the rest of the work was largely done by the Lagonda men. Frank Feeley modernized the shape by getting rid of the high wheel arches and adding a fastback tail. W. O.'s straight-six engine, with its chain-driven double-overhead camshafts and deeper block, provided an enduring source of power and strength. The gearbox, naturally, was by David Brown.

D. B. viewed racing as an important part of the Aston Martin sales strategy and vital for maintaining his cars' performance edge, since 'nothing stresses all the parts of a motor car as much as an actual race'. Dominated by Alfa Romeo and Bugatti in the inter-war years, the Le Mans 24 was the biggest prize of them all, despite not having been contested since 1939. Aston Martin was a minnow by comparison to the big European factories, but if D. B. could assemble enough men with the right stuff he could mount a serious challenge. In 1950 he hired John Wyer to take charge of the racing programme 'for one season only'. A season that lasted over ten years.

Even though Claude Hill had left the company, he regarded the DB2 as the 'pinnacle' of his achievements. It powered Aston Martin to a class victory at Le Mans in 1950 and set the firm on a quest, at all available costs, to win the competition outright. The road-going DB2 was the first proper British GT – grand tourer – loosely speaking, a car that could speed you across Europe and with enough panache to sidle up to the kerbside of a Monaco casino and command respect. Thanks to its racing pedigree the DB2 also brought 105 hp effortlessly to the toe.

Launched in New York in 1950, it met instant acclaim and over 400 were sold, stratospheric by Aston standards. By the end of the production run the three-part radiator pattern had been simplified into the more sophisticated single moustache that set the trend we recognize today. An evolution of the DB2 would also catch the attention of a certain spy novelist who liked to keep his hero outfitted with accoutrements of the highest grade. Let's leave Wyer to grind his teeth while he figures out how to attain the pinnacle of motorsport and follow the spy trail.

Bond's First Aston

A recurring theme in the history of Aston Martin is the continual renewal of energy provided by an injection of stardust at just the right time. We last saw flying Pole Tadek Marek working on a special project for the war effort. He then repurposed the Spitfire's Merlin engine to produce Britain's fastest battle tank, the Cromwell, using spare parts from crashed aircraft. In 1954 he was hired by Aston, and his first task was to reconfigure W. O.'s straight six, which had already grown to 2.9 litres. He went to town and touched up everything from the crankshafts to the cams, timing chains and oil pump to raise the power to 162 bhp.

The car it went into was the final Mark III version of the DB2, released in 1957. The styling by Feely was further refined, but it retained the long bonnet with the main cabin sloping off to the rear. The brakes were upgraded to discs on the front with drums on the rear, but when I had the chance to drive it on the *Classic Car* TV show, neither of them was what you would describe as arresting. The travel of the brake pedal was alarmingly long and with diminishing returns. It felt cruel pushing it that hard, like making your grandad down a tequila on a night out. The wheel felt more reassuring, with the shoulder-width wooden helm that Astons were famous for. The effect the wheel had on the direction of the car was a different matter, and just wobbled the body around on the suspension. That's when I realized I was driving it all wrong.

These cars were built to be manhandled. You rode the brakes hard and threw across an armful of steering to pitch the car's weight. *Then it turned.* After that the steering was more of a guide than a rule. Admittedly the rate of acceleration was tame as the engine burbled away, but on a looser surface it would have come alive. The previous owner knew this.

Former RAF Squadron Leader Philip Cunliffe-Lister had modified this Mark III for rallying. When he wasn't sliding sideways, he often visited friends in Kent who were good acquaintances of Ian Fleming. Fleming spent a lot of time in Kent and by another coincidence the chairman of his father's business had bought Zborowski's home at Higham. Bond almost drove past it on his way to *Goldfinger*'s house at 'the Grange'.

Fleming must have marvelled at the gadgets on Cunliffe-Lister's Mark III: reinforced bumpers, two-way radio connectors, a Halda device for recording speed over distance and a secret compartment inside the centre console for stowing tools. When he wrote *Goldfinger* in 1958, Fleming gave 007 a new pool car which he referred to as the DB3, because nobody can be bothered with the official DB2/4 Mark III, which is far too fussy. Fitted with a homing device, special bumpers and a hidden compartment for a semi-automatic Colt .45, the car had stepped up a gear, at least on paper.

I'm jumping ahead here, but when the screenwriters Paul Dehn and Richard Maibaum set about adapting Fleming's novel for the *Goldfinger* movie in 1963, they included the Aston Martin, complete with all its gadgets, in their story. The next stage was turning words into motion.

Bond's production designer Ken Adam and special effects supervisor John Stears approached Aston Martin, who agreed to lend the production two cars. Tadek Marek was just putting his finishing touches to the DB5 at the time with his all-aluminium 4.0-litre engine.

A series 5 DB4 test car was updated into a fully fledged DB5 and its red paint switched to silver birch to become the gadget car. The film producers exceeded Fleming's accessories with Browning machine guns popping out of the front lights, tyre-cutters extending from the hub caps, smokescreen, oil slick, bullet-proof windscreen, fender-rams and a passenger ejector seat that could be activated from a button on the gear stick.

With the coachwork styled by Carrozzeria Touring Super-leggera of Milan, the DB5 was an instant heart-breaker. The pronounced radiator grille and fenders warned onlookers that it meant business, while the air scoop in the bonnet betrayed the significant 240 bhp lying beneath. Paul McCartney took one look at it on set at Pinewood and ordered one the next day. It cost nearly twice as much as the Jaguar E-Type and was absolutely worth it. You still sat 'on' the Jag, an expression that endured in racing circles until chassis advances permitted the driver to sink lower into the car. The DB5 had state-of-the-art disc brakes, and the suspension was properly supported with springs and an anti-roll bar to control weight transfer so that, finally, the Brits had a Ferrari-slayer.

James Bond powering across the screen in *Goldfinger* carried the DB5 to those dizzying heights that have linked Aston Martin with 007 ever since. The DB5 was the perfect match for Bond's swagger and sophistication.

David Brown loved his DB5 but found that his terrier was a sharp critic of his gear changes. To prevent it snapping at his fingers from the passenger seat, D. B. commissioned a gorgeous estate version – sorry, a 'shooting brake' – of the DB5 so that he could stick his hound in the back. Peter Sellers ordered a customized edition too.

I've had the privilege of driving the DB5 in three Bond movies, so we have become old friends. The first time I climbed aboard was on a chilly night in East London. Lee Morrison, who is now the 007 stunt coordinator, was on hand and dressed like a bomber pilot with multiple layers of goose down and large Sorel boots. Lee could double for a young Ian Fleming with his long features and penetrating eyes, but to annoy him we call him Inspector Gadget. He was a multiple motocross champion as a teenager until he broke his neck and made the switch to stunts. Whatever I'm about to do with the car, it won't impress him.

Bond's oil-stained lockup garage was beautifully lit by Sam Mendes to throw up the shadow lines of the DB5, lying in wait like a smooth grey torpedo. There's something chameleon-like about the car's distinctive silver-birch colouring, always changing with the mood. Daniel Craig was leaning against the wall discussing his scene with Sam but neither man was able to keep his eyes off the Aston. This one first appeared in *Goldeneye*.

Gliding across the dark grey leather seat, worn brown at the edges, was a lot easier than squeezing into the tight Bertelli motor from the 30s. The chair was soft but no sofa. The big three-spoked wheel was canted forward a few degrees into the perfect driving position. The fun wand below the right of the chair was a fiddly flyaway handbrake that I never trusted to be either fully off or on.

No belts to speak of, more fashion statements, and no head rests, which was great for reversing.

Nobody was looking, so I took a sneaky look inside the secret compartment. (Smokescreen buttons, check!) The tiny gear lever had a little cap so I flicked that open and saw the red button for the ejector seat. I was just wondering if it was armed when—

'Ben,' came a voice from the window.

'Yes, er, Lee,' closing the cap fast.

'Come take a look at this, mate.'

The route was simple enough. Bond was taking M out of London up to the safety of his remote estate in the Scottish Highlands. My bit was the quick getaway from the garage, down a lane and right onto the road. Lee helpfully pointed out the blinding set lights that masked the turn. I noted a mark on the wall parallel to the roadside kerb so I could use that to time my run. Clipping that kerb with the wire-spoked wheels would be asking for a P45.

I climbed aboard and hid the radio under my right thigh so it didn't fly around. Popped the engine to let the old girl clear her throat. A strong gust of petrol fumes filled the garage as the triple carburettors made up for the cold. Crew were ushered away and the set was locked down. The cameras got rolling and the stunt channel opened in unison with the set comms. 'Roll video . . . Stand by . . . We're rolling . . . And three, two, one . . . *Action!*'

I dropped the clutch in first, peeled the skinny rubber and took off. The DB5 was ultra-forgiving, slithering ninety degrees easily from the garage and towards the bright lights bouncing off the single-pane windscreen. I plucked second gear, turned on the mark and floored it up the road. The speed bumps looked bigger from inside the car but the DB5 leapt them easily without needing to slow down or worry about ground clearance. Into the final left the body rolled gracefully and the shot was over. Daniel took over from there and the next stop was over the border and into the mist-capped mountains.

I don't mean to sound ungrateful but I only wished that the DB5 had a bit more power. The handling was soft and it could have taken, say, another 100 bhp and become a completely different animal. Tadek Marek agreed but he kept this to himself. In 1964 he was already working on a completely new V8 engine that would power Aston Martin all the way into the 80s.

What is less known is that Tadek sneaked a V8 into a DB5 and used it as his private car from 1965. He widened the front of the chassis to accommodate the bigger lump and installed a De Dion bar that stretched across the rear axle to reduce body roll and handle the power of the 325 bhp motor. Engineering director Dudley Gershon helped with the engine configuration and recalled its 'super torque curve had us drooling, it would idle like a dream and pull beautifully through the range'. With its user-friendly mid-range torque and full-throated roar through twin exhaust pipes, it must have been the perfect blend of style and performance. It was the only one of its kind, and some lucky buyer managed to acquire it in 1969. To drive it would be to dream, but one day we might get close.

Red Mist

The 50s was a strange and wonderful time for Aston Martin, when everyone forgot that the aim of the business was to sell cars. After the DB2, Astons raced ever more advanced machinery that was largely unavailable to punters. At one point the racing department was running five different models in the same race while simultaneously developing a Formula 1 project and competing with the production side for resources.

The two breeds of racing had become quite distinct, with F1 denoting open-wheel, single-seat racing cars competing in short sprints, while the two-seater sports cars retained a closer relation with the road-going machinery, being capable of longer distances and with their wheels enclosed by bodywork. This gave drivers the opportunity to run two careers simultaneously and with a certain amount of flexibility, since sportscar racing required two drivers to share the innings. Ferrari competed in both categories.

Astons scored plenty of landmark victories in sportscar racing's sub-classes but they would need yet more quality to claim outright victory. David Brown had only intended to go racing for one season but something strange happens to grown businessmen when they hear that 'certain sound', the wail of pistons beating a mesmerizing tune that shakes the air and weakens the knees. There was no sound sweeter than the ear-splitting roar coming from the tail pipes of a Ferrari V12. D. B. saw red and commissioned a project to develop his own V12 to compete with the big boys and shoot for the moon.

Professor Eberan von Eberhorst had worked for Dr Porsche at the advanced Auto Union racing team and joined Astons as chief engineer. Accustomed to big budgets, he was constrained at Astons, where cutting corners was the only way to get things done. It flew in the face of his need for precision. During the war he had improved the fuel efficiency of V-2 rockets, extending their killing range impressively into the distant reaches of Britain, and this perhaps didn't endear him to the lads in the canteen over tea and biscuits. He was so painfully methodical that the DB3 he began designing in 1950 never actually drove off his drawing board.

Willie Watson was the polar opposite and couldn't sit still. He took von Eberhorst's plans and wrangled them into the beautiful DB3S racer, which proved to be a class winner. Willie was one of W. O.'s men from Lagonda, where the pair had designed the LB6 engine that was still being flogged in the race cars. Like the Incredible Hulk it had grown from 2 to 3 litres and been juiced with Weber carburettors, but it was out of date and underpowered. Willie was tasked with building a V12 to double the number of pistons powering the motor and take the fight to Ferrari.

Willie escaped the vigilant eyes of the more prudent von Eberhorst and knocked up something that resembled a double whammy of the LB6. He ditched its cast-iron crankcase and instead used an aluminium case that expanded at a different rate from the spinning internals, causing mayhem with oil pressure and load. On its first test drive, by David Brown of all people, it caught fire.

For Le Mans in 1954, John Wyer had succeeded in removing the team from the noisy Hotel Moderne in the town centre and installing them at the more remote Hotel de France in La Chartre-sur-le-Loir. Nestled among vineyards and bordering the river, this sleepy French village had no idea what was about to hit it.

Peter Collins was a rising Formula 1 star who had been seconded to Astons for sportscar events. A debonair playboy and speed demon, he took the circuitous route to La Chartre, his blond locks billowing in the air as he sped through Le Mans town. A gendarme was too slow off his mark to intervene, and Peter offered him the traditional Agincourt two-fingered salute, hooting with laughter until he reached the second officer lying in wait for him. His mastery of French improved rapidly as he faced *le grand homme avec baton*. Roy Salvadori was another seasoned British F1 racer and an Aston stalwart. Tall, dark and handsome, he was constantly mistaken for an Italian and was a hit with the ladies. Some of his conquests later realized that their initial delight in his qualities as an attentive listener had been misplaced, when they discovered that he was deaf in one ear. Reg Parnell rounded off the ringleaders of the merry band of ten drivers piloting five cars of varying shapes and sizes. A burly F1 hero in the twilight of his career, Reg bent cars to his will through gritted teeth.

The Hotel de France's art deco façade overlooked a quaint little square. The armada of Astons arrived en masse and turned it into a car park, with the principal machines lined up outside the bar-cum-restaurant on the ground floor. At first Madame Pasteau, the ageing hostess, had not believed Wyer when he said he wanted to book all forty-five rooms. She presumed that yet another Englishman had lost his mind. Once she had been assured of his sincerity the team adopted the hotel as their home, though some of the amenities were less than familiar.

Wyer's bodywork specialist mistook the bidet for a foot-washing device and put his foot straight through it, causing himself some injury. Collins meanwhile was adjusting to the cuisine and found all the bread a tad stifling in the stool department. After siring a fine specimen in the communal loo, he attempted to flush it away but the cistern was on strike. Despite the engineering genius on hand, nobody had a solution. Poor Peter had no choice but to seek out Madame.

The old lady climbed the stairs and warily entered the cubicle with a small crowd of Peter's technical advisers. Using mime, since she spoke little English, Peter first directed her attention towards the substantial offering on display in the pan. Then, with the theatrics of a circus ringmaster, he exclaimed, 'Regarde!' and pulled heartily on the chain. To his amazement, but not hers, the toilet flushed instantly and the turd vanished. Madame scoffed and left, never more convinced of the imbecility of the English.

D. B. flew across in his private plane and must have felt that if this was a game of roulette, he had good coverage across the table with each of his machines driven by a pair of highly skilled drivers. The DB3S race cars with their stunning elongated wheel arches had roofs attached and new twin spark plugs to boost their power by 40 bhp, making them both slippery and fast. A sister car had been supercharged, and the new V12 was running as a Lagonda to teach Ferrari a lesson.

It was hoped the Lagonda V12 would produce 350 bhp, but it only managed 312. It was slow on the straights and bled oil before it was crashed within two hours of the start. Both DB3S coupes had aerodynamic defects that revealed themselves at high speed in the form of rear lift, which was the opposite of helpful. They crashed within yards of each other at the ultra-fast White House hump. The supercharged DB3S hung on the longest, but the engine was showing signs of overheating and was losing water pressure. Reg Parnell nursed the throttle to keep the gauges in the green until Salvadori climbed in, turned his deaf ear to the engine's protests and kept it wide open until it blew a gasket.

The polished green metal fleet that had glinted in the sunshine just a day earlier was scrunched into a pile of scrap in the corner of the paddock. The scale of the multi-faceted operation, continuous late nights and catastrophic failure were too much even for the stoical Wyer, who felt 'the return from Le Mans was a retreat from Moscow'. He wanted to throw in the towel on the rest of the season.

'I don't agree,' Brown said. 'This is the time when you get back into the race as quickly as you can.'

'With what?'

'That's your problem. But I do insist,' D. B. replied. 'You can't just retire hurt.'

Wyer was given some time off as he was on the verge of a nervous breakdown. On his return he commissioned an independent review of the V12, which revealed fundamental faults in its architecture. The clearance around the crankshaft bearings was so tight that it couldn't be started without being preheated, and when the engine ran up to temperature there was uneven expansion that caused oil to drop through the gaps. In short, the V12 in which so much had been invested was a lemon that would be buried 'unloved and unmourned'. If Aston Martin wanted to mount a serious campaign they needed to find speed *and* reliability. But at least David Brown wasn't the only team boss having a nightmare.

Mossa!

Watch him – he will be one of the greats.
Tazio Nuvolari on Stirling Moss, 1949

The blood-red car scorched across the finish line in a dominant flag-to-flag victory. The boy wonder, barely twenty-one years old, made it dance like it was an extension of his body. His style was clean and dirty at the same time. It had parallels with Enzo Ferrari's hero, the great Nuvolari, who had wowed the *tifosi* – the irrepressible Italian fans – with his four-wheel drifts through the curves. Nuvolari was a fading memory. The Ferrari fans only cheered one name now, and they chanted it in their own style: 'MOSSA . . . MOSSA . . .'

'Mossa!' he cried out. Suddenly Enzo Ferrari was awake, and the pit of his stomach relaxed. It was just a bad dream.

He cleared his throat and spat. Checked his watch. There was still time. He got up and went straight to the telephone, made himself absolutely clear and ended the call.

Stirling Moss's father was a dentist and just happened to be studying in Indianapolis when the Indy 500 was held in 1924. Armed with a letter from a Mercedes dealership that said he could drive, Alfred Moss blagged a seat in one of Henry's Fords at the fastest race on earth. He finished a respectable sixteenth but returned to the more profitable and safer world of dentistry.

Alfred noticed the special qualities in his son, though he was blissfully unaware of the daily beatings Stirling was taking from other boys at school under the pretext of him having a Jewish grandfather. Rather than await divine intervention, he learned to box and to stand his ground. Stirling was equally good riding a horse, and won over a hundred show-jumping prizes. But cars offered a purer expression of his ability.

Moss first sat in a competition car when he was seventeen and won the event. By 1950 he had won enough races in his little Cooper single-seater, and every other car he could get his hands on, to attract the attention of John Heath's professional Formula 2 team. Heath entered Stirling into the Grand Prix at Bari, where he humbled the drivers of several more powerful Formula 1 machines, placing third behind Giuseppe Farina and Juan Manuel Fangio in their dominant Alfa Romeos.

Farina had barged past Stirling during the race but overcooked it and was instantly re-passed by the upstart. Farina composed himself for a cleaner overtake, passed him and gesticulated in the Italian style. Fangio followed, 'laughing his head off' at Farina's humiliation. 'Fangio endeared himself to me from that moment onwards,' recalled Stirling.

Enzo Ferrari had noticed Stirling's prodigious talent and wanted a meeting. 'When you're twenty years old and you're asked to go to meet Ferrari you just cross yourself and face Modena,'

Moss recalled. The Mosses flew to Rome and then endured an uncomfortable train journey to the sunbaked port city of Bari, where Ferrari had assembled his test team. There was no sign of Enzo and nobody acknowledged them. The mechanics carried on about their business, so Stirling wandered into the garage and climbed into the car.

'What are you doing?' Ferrari's engineer demanded.

'I'm Stirling Moss and I'm driving this.'

'No you're not, Piero Taruffi is.'

Enzo was famed for playing mind games, and used rivalry to put pressure on his drivers to perform beyond their natural limits. He was also a jealous man – jealous that no driver should take credit for his cars' successes away from him. Denying Moss the drive was intended to put the youngster in his place. Ferrari's trick backfired spectacularly. From then on, Stirling Moss only wanted to beat those red cars.

Alfred Moss had manufactured bomb shelters during the war, and Stirling was fiercely patriotic. He spent his early years in F1 driving less competitive British cars until Mercedes and then Maserati took notice and picked him to drive alongside 'the greatest F1 driver of all time', Juan Manuel Fangio. Their relationship was akin to that of father and son or a master with his apprentice. Fangio won a record five world titles during the 1950s. He had supreme pace, natural control over the slithering brutish cars of that era and an astute awareness of risk. He avoided unnecessary skirmishes on the track and was loved by those who knew him.

At the Spa GP 1953, Maserati fielded four cars. The maestro

was fastest, and one of his teammates, the Anglo-Belgian Johnny Claes, asked if he could try his car as he felt his own might be slower. Fangio agreed, and Claes tried it but could go no faster, so he asked Fangio, 'How on earth do you do it?' The answer was profound: 'Less brakes, and more accelerator.' Braking when entering a corner was, and remains, the most artful element of driving, and Fangio explained to Claes that he needed to release the brake earlier. Show me a modern F1 driver who helps a team-mate and I'll eat my flameproof socks.

In the days before data logging revealed every secret to a driver's lap time, Fangio was remarkably candid about his method, right down to which gear he used at every corner. He knew he couldn't be bettered, until one day he woke up with a headache.

'We have a phrase in my country, *mucho dolor de cabeza*,' Fangio explained. 'It means, roughly, a big headache.' When Fangio saw Stirling climb aboard his first competitive car at Vanwall, he reached for the aspirin. 'Of the five men I rate the best racing drivers in the world, they are Stirling Moss, Mike Hawthorn, Peter Collins, Tony Brooks and Jean Behra,' and 'Moss, he is the greatest of them all – and I include myself.'

Luckily for Fangio, Moss ended up as his team-mate, accepting the unspoken rule that he was number two to the great man, and he was happy to follow in his wheel tracks. Stirling was fair to a fault as well as loyal.

Changing the Guard

When Mercedes withdrew from motorsport in 1955, Stirling became a free agent for sportscar racing. He considered joining Jaguar since they were British and had just won Le Mans. The problem there was that the team manager, Lofty England, was misty-eyed for Mike Hawthorn and had signed him as number-one driver. Stirling could bestow that accolade upon Fangio, but nobody else was getting it.

Lance Macklin had been driving for Astons since 1949 and introduced Stirling to John Wyer to discuss joining the race programme. Lance waited in the car park until Stirling returned looking hot under the collar. The proposed retainer – £50 – was an insult and he had blown his top.

'That's what I get,' Lance explained.

'You must be out of your bloody mind,' Stirling exclaimed. But it seemed that he had misheard – the retainer on offer was actually £1750. But he must have been impressed enough by Wyer because he signed to Aston Martin anyway. The point on money was more about respect because in those days, drivers supplemented their salary with sponsorship deals, notably from oil companies, and had parallel careers that paid the bills. And besides that, Astons had something the other teams didn't: they raced for the craic.

Stirling went to Goodwood for his first test. Mechanics are a loyal bunch, so they didn't warm instantly to the big gun from F1 muscling into their little team. Stirling drove the little DB3S onto the track and was on it immediately, assessing its handling characteristics: solid brakes, easy handling, underpowered and not enough front grip. It had a tendency to understeer at the limit so the steering didn't point him precisely where he wanted it to.

He brought the car in and asked the boys to reduce the air pressure in his front right tyre by 2 lb. It was a subtle change but on a clockwise circuit like Goodwood with a lot of long right-handers it could make a difference. The mechanics rolled their eyes at each other and decided to see if His Royal Highness would notice the pea under the mattress. They held the tyre gauge to the wheel but without adjusting the pressure and sent him out. Moss came straight back in and said, 'The tyres are still the same.' They had been caught out by the maestro. Their respect for Moss grew from there.

The lead drivers at Astons were Reg Parnell, Peter Collins and Roy Salvadori, with Tony Brooks in the ascendant – and the pecking order was about to be put to the test. Wyer had colour-coded the cars with yellow, red, blue and green markings around their radiator grilles to help identify them, a handy trick for timing and urgent pit calls. Occasionally, when a driver complained, 'His car's better than mine,' Wyer swapped the colours round to see if they even noticed they were driving a different machine. But you had to get up pretty early in the morning to put one over on old Reg, who spotted the wet paint.

The DB3S used Girling disc brakes with friction that increased with heat and use, requiring less foot pressure to slow the car the further you drove. Drum brakes were harder underfoot and faded with heat, but Peter Collins preferred them. He had also discovered that the unsprung weight of the discs worsened the car's handling, so his DB3S was the only one that was fitted with drums for the race at Rouen in 1956.

Stirling was fastest in practice but he noticed something better about Peter's car, so he tried it and went even quicker. As the number-one driver he could drive whichever car he pleased but he asked for Peter's permission, which Collins duly gave. But Peter was hot stuff with Ferrari in F1 at the time and once he had time to think about this arrangement, it stuck in his craw.

Wyer always brought the drivers in early for a pre-race briefing. He would explain the format of the race and always emphasized the rev limit, 6200 max. Abuse of these engines was not permitted because once you pushed beyond the red line it was like playing Russian roulette with the pistons.

The mercurial Peter had been known to say, 'I don't care what the team orders are, I'm out to try and win the race,' and he was having one of those days. As the race got under way Peter shot to the front with his rev needle buried firmly in the red zone. After a few lusty laps entertaining the crowd with some ballsy slides and steamy braking, his engine threw up the white flag and he came into the pits. Wyer approached the car, which was radiating heat, and flicked the telltale on the tacho. It read 7300 rpm! He couldn't blame him. Pete was a fighter, but his future lay with Ferrari.

DBR1

By the mid-50s, everyone at Astons was fed up with the domination of the Italian teams. It was high time to develop a thoroughbred racing car with none of the drawbacks inherent in overstrained road-car technology. The LB6 engine powering the DB3S was at the end of its life so Wyer, who had been promoted to technical director, appointed Ted Cutting to oversee an entirely new design from the ground up in 1955.

Ted was a senior draughtsman at Astons and, having lent his expertise to the successful DB2 chassis, was the right man to carry the baby. Nearly every component was drawn from new, including a modern space-frame perimeter chassis that reduced weight and lowered the centre of gravity. The wheelbase was lengthened, and the proven front suspension and steering mechanism was brought over from the DB3S. The body was made from toughened aircraft aluminium sheets shaped on a wheel and provided a thin but strong, sculpted shell. Lighter Girling disc brakes reduced the issue of the unsprung weight inside the wheels throwing the tyres off the tarmac.

The prototype was put in a wind tunnel where it developed slippery aerodynamics, but retained some foibles. Air pressure built up under the bonnet, which had to be tethered down to prevent it going into orbit. During testing on the high-speed bowl at MIRA, the vehicle development and proving ground, the driver felt like he was being strangled and it was realized that a vacuum was being created inside the cockpit, so the windscreen was adjusted.

Ted Cutting's 3-litre RB6 engine featured an all-alloy crankcase, leaving behind the barrel type that had plagued the old straight six and notably the doubled version in the V12. With twin chain-driven overhead camshafts and two valves per cylinder, it efficiently produced 250 bhp at 6000 rpm. The whole car weighed

1760 lb, so this was an excellent power-to-weight ratio providing a maximum speed of 152 mph.

The only element that escaped Ted's complete control was the gearbox, for financial as well as political reasons. Mercedes had spent a fortune on their racing programmes, and their gearboxes reflected years of advanced and heavily researched engineering. To save money, Ted 'acquired' the drawings of the gear mechanism from a Mercedes Grand Prix car 'through British intelligence'. He based his outer casing on a Maserati's. The parts were manufactured by David Brown Engineering, but since they were used to making heavy-duty commercial gears, that is what they produced. Ironically, for a team owned by a gear manufacturer, this proved to be the DBR1's Achilles heel. Shifting gear was laborious and, under pressure, broke down altogether.

The new race car was unlike anything that had rolled out of the Feltham factory before. The surly grille of the nose hunkered down low and was flanked by long curvaceous front-wheel arches with graceful lines that curved up and blended seamlessly into the side profile alongside the cockpit, then up over the rear arches and into a teardrop at the tail. The long smooth bonnet extended to a small windscreen that flicked the air over the driver's head, behind which a head rest and curved hump helped to cut through the airstream like a bullet. Dressed in deep British racing green, she was truly deserving of her unique moniker: DBR1.

Green Hell

When car manufacturers want to find out what will break their cars they hand the keys to a racing driver and send them round the Nürburgring. The circuit is a 17.5-mile loop that twists and winds through the Eifel Mountains in western Germany. Whereas a normal track might have one or two corners where you need to tighten your groin straps, the Ring has trouser-soiling turns everywhere.

In 2001 I raced in the World Sportscar Championship in a 200 mph Le Mans prototype-class car called an Ascari, taking on the Ferrari 333s with their screaming V12s. 11,000 rpm. Music. We were racing on the short track that F1 currently use and got bored with that, so we hired a pair of bog-standard cars from Avis, peeled off the stickers and paid ten euros at the gate for a crack at the full circuit.

My stocky South African team-mate Werner Lupberger was the ice man with his buzz haircut and wraparound sunglasses. A man of few words beyond the topics of BBQs and racing, he lined up alongside in his identical Opel, gurned at me and floored it. For once I was glad that he was in front and encouraged him with some helpful taps on the bumper. The track width was generous, and we could see where we were going at first, but the sense of uneasiness built as our speeds rose towards 100 mph. Many of the heavy braking zones were well telegraphed, but the never-ending sequence of very fast, often connected corners blended into one, and it was easy to mix them up and fly off. Werner was so bull-headed he wouldn't have cared.

Wherever you looked it was the same type of pine tree zipping past, the same metal barriers, same red and white kerbing in the corners, same bumps, same grass and the same dark tarmac except one golden patch around the carousel. I saw this loop coming on the satnav and sneaked past Werner. From then on the track narrowed slightly and took on a more aggressive personality until the colossal back straight, where we tapped out in the Opels for what seemed like an eternity. We returned the smouldering cars to Avis, and Werner complained at the drop-off that the brakes on his were terrible.

As John Wyer packed his cars off to the Ring in 1957, his ears were ringing with the predictions of Professor von Eberhorst, who had warned him never to go there: 'You'll destroy your cars, you'll never finish . . . ' Oh, Eeyore.

Learning the Ring was like memorizing the Bible. Even drivers who had raced there numerous times would barrel up to a corner and suddenly ask themselves, *Where am I?* It was even harder in the 50s because there were hedges and grass banks masking the corners. Tony Brooks was a dentist as well as an F1 driver and applied his methodical, medical approach to the task. Using an article by Paul Frère about his experiences at the Ring, Brooks noted the gear-change points on a map. He memorized the key landmarks and used these to split the circuit into sections so that if he got lost he could quickly reorientate himself. Wyer saw that Brooks was ahead of the game and ordered Roy and Peter Walker to go with him as passengers.

The mild-mannered Brooks, with his gangly physique and prominent gnashers, was not the archetypal daredevil to look at, but he was a blisteringly fast driver and exceptionally smooth. Fuel consumption is usually married to a driver's lap times, and the faster you go, the more you accelerate and so the more fuel you burn. The same is true with tyres and brakes because driving harder tends to create more friction and heat. Yet Tony used less fuel and consumed fewer tyres and brakes than any driver on the team. He reckoned, 'Any steering correction you have to apply is slowing you down.'

Tony took to the course, and the boys clung on to whatever they could as the scenery began to blur. Brooks provided an urbane commentary like this was some routine surgical procedure, while the view from the back seat was the wrong end of the microscope as far as Walker was concerned. As they flew over the crest at Pflanzgarten, the engine whinnied, Walker's head hit the ceiling, then his arse felt the clash of metal as the car landed. Enough was enough. When they pulled in at the end of the lap, Walker dived out of the car and ran off in search of a brandy.

Unfair Advantage

In 1957 Enzo Ferrari had sacked his team manager and installed a more compliant operator in Romolo Tavoni. Tavoni didn't know much of anything about motor racing but he did the two things Enzo expected of him, which were to answer the phone and follow orders.

For the Nürburgring Maserati brought out their big guns with Fangio and Moss driving the bestial 450S with its 400 bhp 4.5-litre V8. Moss was only contracted for selected races with Astons and was driving as many as eight different types of car per season. In practice Moss posted the fastest time of 9 minutes 43.5 seconds, with Fangio just one second slower. Brooks in the DBR1 managed a time of 9' 48.2", which was nearly forty seconds faster than the old car had managed and considerably quicker than anyone else in the team.

Pandemonium reigned at Ferrari. The drivers were swapping cars round, and a confused Wolfgang von Trips mistook the centre accelerator pedal in the Berlinetta for the brake, shunted heavily and broke a vertebra. Even with Collins and Hawthorn at the wheel, Ferrari struggled to break the ten-minute barrier.

At Astons Reg Parnell, having hung up his helmet, was now Wyer's right-hand man and running operations. The poacher turned gamekeeper was proving to be an astute team manager. Hotshoe Roy Salvadori was still lapping considerably slower than Brooks, but the comparatively inexperienced Noel Cunningham-Reid was showing promise, so Reg switched him over to partner Brooks.

The start of the race was conducted Le Mans-style with the glittering ensemble of green, red, silver and white cars lined up diagonally along the pit wall and the drivers facing them from the opposite side of the track. As the starter's flag fell, there was silence except for the pitter-patter of leather soles as the drivers scampered across thirty yards of concrete. They leaped aboard, fired up the engines and shot away. There were still no belts, the logic being that in the event of a crash the cars would normally catch fire so you needed to get out quickly.

Moss was first in. His car started forward but then stalled. Hawthorn heard his Ferrari roar to life, let out the clutch and realized he had mistaken Peter Collins' engine alongside him for his own and went nowhere. But all this was in Brooks' rear-view mirror as he gunned his DBR1 towards the most treasured view to a racing driver – empty track.

Moss got going and joined the fray in a distant fifth place, and proceeded to break the lap record time and again until he over-hauled Brooks, who couldn't counter the Maserati's straight-line advantage. But when Moss lost a wheel, Brooks regained the lead.

To the astonishment of all, young Cunningham-Reid then extended the lead given him by his co-driver Brooks before handing it back and staggering into the pits. The mental and physical exertion demanded by the Ring was so acute that he spent a long time alone afterwards with his head in his hands, unable to speak.

When Fangio's car sprang an oil leak that was both big Maseratis done for, although Moss, with his never-say-die atti-tude, spent the rest of the day hopping from one spare car to the next. Nevertheless, Brooks snatched the chequered flag in the first British car in history to win an international competition at the Ring. Ferrari simply could not match the Astons for pace and had been beaten into submission, though they still collected a good points haul in their march towards that season's World Sportscar Championship title.

In Grand Prix racing Brooks had been overshadowed by his team-mate at Vanwall, Stirling Moss. Brooks lacked Moss's killer instinct, a trait in a dentist for which the rest of us can be grateful. But the Nürburgring had demonstrated his true driving ability for all to see.

At the end of the race the DBR1's engine was spotless but the mechanics had a fright when they removed the bodywork. The chassis tubes were pointing all over the place and most of the welds had broken. Aston mechanic Tug Wilson shouted, 'Put the cover back on for Christ's sake,' before anyone saw it. If the chassis had collapsed the result would have been a catastrophic loss of control. Had it not been for Tony's silky driving style putting less force through the DBR1, the car would never have finished, let alone won.

The FIA officials who governed the sport got together and decided to meddle with the regulations. Alarmed by the high speeds of the cars, they voted to restrict engine size to 3 litres for the following season in 1958. That would suit Astons very nicely and remove the handicap they still suffered against the brute force of Scuderia Ferrari and Officine Maserati with their higher-capacity motors. David Brown decided to focus entirely on sportscar racing and shelved his nascent plans to compete in Formula 1.

In 1958 Aston Martin would finally have a full shot at the World Sportscar title and an outright win at Le Mans. Stirling Moss was signed for the whole season and the volume was turned up to 11.

Testa Rossa

Enzo Ferrari had caught wind of the 1958 rule changes early and began developing his 3-litre V12 engine for the 250GT as early as April 1957. The secret weapon of his engineers at Maranello was a radical body design by Sergio Scaglietti. Its menacing self-contained frontal air scoop extended forwards like an aardvark and was attached by thin strakes to the shark-nosed wheel arches at its shoulders. The pronounced features were largely intended to resolve cooling issues for the Ferrari's drum brakes. Enzo still didn't trust discs. Scaglietti jokingly called the new Testa Rossa a 'Formula 1 car with mudguards'.

Ferrari's V12 engine was a tried and tested campaigner originally crafted by Gioacchino Colombo but enhanced by the ingenious Carlo Chiti, the man who persuaded Enzo to manufacture mid-engine sports cars. Chiti managed to squeeze 50 hp more out of his 3 litres than Ted Cutting's straight-six Aston could find, and the Ferrari could rev safely to 7000. Chiti quenched its thirst by doubling the number of carburettors and used smaller valve springs that allowed him to strengthen the cylinder heads and make the system more reliable.

The Testa Rossa took its name – Red Head – from the red-painted cam covers on its engine, and, true to its name, it took first blood in the opening rounds of the 1958 World Sportscar Championship. Aston Martin skipped the opener in Buenos Aires and then suffered a depressing run of gearbox failures. The next round was the Targa Florio in Sicily.

The Targa was a forty-five-mile driver's delight up and down a craggy mountain. It was a street circuit, Sicilian style, which blasted through villages packed with spectators, nipping in and out of tight passages then opening into blind sweepers bordered by hair-raising drops. There were Monaco-style switchbacks, lots of bumps, abrupt kerbs, luscious trees and pretty girls.

To be competitive on a street track was a balancing act between attacking fearlessly as the walls flashed by and accuracy to avoid ripping the wheels off. Drivers likened the mindset to leaving their brains behind when they stepped into the car, and Ferrari's lead driver Luigi Musso had more on his mind than the rest of the grid combined. Musso had significant gambling debts, and he needed the prize money from F1 and sportscar racing to pay them off. His team-mates Mike Hawthorn and Peter Collins had also made a pact to split their prize winnings and work against him. This was an unprecedented arrangement in Formula 1 where the very nature of single-seater racing, with one driver per car, was geared towards single combat as opposed to the more collaborative endurance sportscar races where co-drivers took it in turns and shared a machine. To complicate things further for Musso, he had left his wife and children to go off with brunette bombshell Fiamma Breschi.

Due to the restricted track the racers were sent off individually at forty-second intervals with their starting positions chosen by lot. Musso started last of the main contestants in his Testa Rossa, while his rival and team-mate Collins started at the front. By the end of the first lap, Musso had torn through the field to take the lead. Stirling made an uncharacteristic mistake and clobbered a marker stone, losing over a minute, then was delayed again by a mysterious vibration. When all was patched up, he set off in full attack mode. Jean Behra was driving a small-engine Porsche which he spun, missed everything and then continued like a terrier.

It was clear to those who knew him that something special was stirring in Moss. Dennis 'the Beard' Jenkinson had been Stirling's co-driver for their record-breaking run with the Mercedes 300 SLR at the Mille Miglia in 1955, a 1000-mile course that made learning the Nürburgring seem like a nursery rhyme. The Beard had read a six-metre reel of notes from a dash-mounted 'toilet roll' and used hand signals to indicate the route to Stirling, who responded unflinchingly. Stirling thought, 'Jenks did not understand what danger was. I think to be a co-driver in the Mille Miglia you need to be really weird.' Dennis was at the Targa as a journalist and noted the way Moss took the 'pit curve in one long seemingly incontrollable slide and roared off into the mountains'. He shaved a full minute off his own Mercedes lap record, *even with a spin*, and was due to hand over to Tony Brooks when the Aston's infernal gearbox packed in.

Musso drove a flawless race to win the event despite losing his brakes on the final run. Behra had muscled his way past Hawthorn and Collins to claim second. The points standings put Ferrari into an unassailable lead in the championship tables. However, nothing counted with Enzo more than winning Le Mans.

As the dust settled on Sicily, Moss was kicking himself. Mistakes had been made. His desire to go further and faster than any other driver burned deep.

Redemption

If everything is under control you are just not driving fast enough.

Stirling Moss

When Cutting designed the DBR1 he distributed its weight as evenly as possible, with the engine in the front and the transverse gearbox mounted across the rear axle. By bringing the weight marginally closer towards the DBR1's centre, nearer the driver, he raised the car's polar inertia, making it keener to rotate into corners.

For a driver who was willing to hang the tail out this meant that you could aim towards the corner and drift through it sideways, making slight adjustments with the steering to track direction. The trick with sports cars was the bulky wheel arches, which obstructed the view of the corner as you got closer to it, exacerbated by the cars' long bonnets. That was partly why Fangio preferred Grand Prix cars – he could see his wheels. Sportscar drivers had to be pitch perfect with each four-wheel drift, so as to slice the nose, which they couldn't see, tight past the barriers to take the maximum return on the extended radius of the curve. It took guts as well as skill to throw the car towards something you might crash into.

Astons sent three cars to Germany in 1958 with the Ring-meister Tony Brooks sharing his DBR1 with Lewis-Evans, Jack Brabham joining Moss, and Roy Salvadori paired with a man wearing a Stetson and cowboy boots. Carroll 'Shel' Shelby was a chicken-farmer from Texas and had the blue and white striped overalls to prove it, preferring them to the stuffy racing onesies. Carroll had cleaned up in American racing in the SCCA series, which is where Wyer spotted him.

He was 6 feet 3 inches tall and spoke with a Southern drawl unintelligible to the Brits until they got used to it. They were enthralled by Shelby's tales of money and descriptions of his farm machines that spat out dollars with every *ker-chunk*. In reality he had been broke since his chickens were wiped out by bird flu, but he was a natural engineer and a good foil to Roy's less mechanically sympathetic approach.

Scuderia Ferrari were running true to form with nobody knowing which car they would be driving, or with whom. After assessing their pace in qualifying, Tavoni phoned Enzo and put Collins with Hawthorne, Phil Hill with Musso, Seidel with Munaron and Von Trips with Gendebien. The result of all the car swapping was that Musso's fastest time was credited to the car, not himself, so he was demoted to seventh on the grid.

Hawthorn had really opened the tap during qualifying to claim pole position in a marked improvement on his performance from the previous year. Moss was fastest of the Astons, nearly five seconds quicker than Brooks, but over two seconds slower than Hawthorn's Ferrari. However, Ferrari had probably pushed their engines harder in qualifying than they could risk during the race.

Moss requested shorter gear ratios, which would give him higher revs to increase his acceleration out of corners, albeit at the expense of top speed as the top gear would run out of legs.

It seemed an unlikely request given the three-kilometre straight, where he would be eaten alive by the Ferraris, and the thought of overrevving gave Reg Parnell palpitations. Reg reluctantly agreed, although cannily only went halfway towards what Moss wanted.

In a scene reminiscent of his first test at Goodwood, Moss tried the car and told Reg that the engine must be knackered.

'Why, what's wrong?' he asked innocently.

'I know you dropped the ratio, but as I'm only getting 6300 rpm when I should be getting 6500, obviously the engine's off-tune.'

Reg realized he had been made. He changed the drive to the shorter gearing and vowed not to say a word to Wyer in case Moss blew it up in the race.

The drivers faced up for the sprint to their cars, all in the certain knowledge that one man would always reach his first. Moss trained relentlessly and could run the 100 yards in ten seconds flat. He was invincible. The starter prepared his flag, but before it fell, Hawthorn was off.

'You bastard, Hawthorn,' cursed Moss, glued to the spot and as ever playing a straight bat.

Hawthorn reached his car, dived in but was paralysed with laughter and couldn't start it. Moss and Brooks tore off in their Green Astons and showed a clean pair of heels to the rest. Stirling proceeded to break the Ring record with every lap until handing the baby over to Jack Brabham. It was Jack's first weekend in the DBR1 and he had only managed six rounds of the Ring in practice. Burdened with Reg's promise of a fate worse than death if he buzzed the engine or threw the car off, he took to the track cautiously and was reeled in by Hawthorn.

In the third Aston pairing, Roy Salvadori was having gearbox problems, again, while Brooks had a spin that resulted in an engine blowback and a small fire. After being doused by a helpful spectator, Brooks returned to the fray. Moss took the car back from Brabham in second place and closed the deficit on Ferrari, shooting past when Hawthorn's handover to Collins was held up by a tyre jack. But the Ferrari pairing was too hot for the lead Aston, and after the final pit stop Hawthorn had Moss in his sights again. On his way past the Aston crew, Hawthorn flipped the V at Reg, who grunted and smiled back.

Stirling drove a series of metronomically flawless laps, even passing Tony Brooks to put him a whole lap behind, while in his impossible bid to close the gap Hawthorn flew off the track at the Carousel. Moss won and crossed the line feeling 'completely dead-beat and quite ill' having driven 36 of the 44 laps. His pulse remained at 130 bpm for a considerable time afterwards. He performed the formalities demanded of the victor but when he got home it took him a week to recover fully. It was a weakness he planned to remedy.

The result put a fair dent in Ferrari's confidence, and you can imagine the hand-wringing that ensued. Maranello responded by fielding an armada of eleven works and privately entered cars for the 1958 Le Mans 24 Hours. Jaguar were represented by private entries of their new 3-litre motor in the D-Type.

Astons pitched up with the same three-car line-up, minus Shelby, who was off sick. Moss was still experimenting with short gear ratios and was trusted to lift off the throttle on the straights to respect the rev limits. The DBR1 was becoming an oversteering car, which suited Moss, who drove by the seat of his pants. The higher-revving gears improved his car's stability by squatting the rear down in the high-speed corners, Moss recording that he could take the first corner almost flat out.

The Ferraris with their drum brakes were no match for the Astons under braking, and Moss stuck his car on pole with Tony Brooks just a second slower, the fastest Ferrari driven by Hawthorn some way adrift. This looked like being the year when Aston Martin would finally stake its claim to the greatest race in the world . . . If Lionel Martin was looking on, he had everything crossed.

Moss ran away with it from the off, pulling away from the fast-starting Hawthorn until he became a distant memory in the mirrors, the DBR1 responding to his will like a well-worn glove. With its assured strength under braking, Stirling probed the pedal deeper each lap. The exhaust of the engine was back-popping and snorting as he bled the throttle free into the fastest corner of the track at Indianapolis and slithered into Arnage.

'It took me eight years to achieve complete concentration for any length of time – the most difficult time to keep concentrating is if you are leading or losing by a wide margin,' recalled Moss.

He applied himself fully to the mission. By the end of the second hour he had eighty-four seconds on Hawthorn and was lapping the entire field as far as third place. Everything was going according to plan. That was, until he heard a loud bang – the sound of the connecting rod to a piston grenading through the side of the engine block. Moss was out.

A terrific downpour offered the opportunity for Brooks or Salvadori to carry on the fight, but their over-pressured Avon tyres provided no grip and they couldn't live with the Ferraris until this was fixed. Before that happened, Lewis-Evans crashed the Salvadori car going through Dunlop Bridge, and some time later Brooks' gearbox broke.

Hawthorn had fried his Ferrari's clutch at the start of the race and it developed problems that put paid to his pace. It was Phil Hill out of California who proved to be the rain master, putting a lap on his closest rival and revelling in conditions where others failed to even stay on the track. Only twenty of the original fifty-five cars made it to the finish line, and in a worrying omen for Astons, Phil found the Testa Rossa 'very driveable' in the wet. He brought the big race home for Ferrari, along with the championship.

The Aston boys packed away their kit and made their way back to La Chartre. When a team has worked all week, all *year* long, and snatched defeat from the jaws of victory, there's really only one thing left to do.

One of the gaggle of mechanics had a bright idea, 'Dirty beer?'

They wandered across to the bar opposite the hotel, Le Cheval Blanc. Glum faces warmed up a bit. A couple of drivers ambled over, then Wyer, then the rest of the gang. Some cheap French wine brought on cheap jokes about their gearbox being made of chocolate, Enzo planting that con rod in Moss's engine and the fact that Roy would never finish a Le Mans if his team-mates always crashed.

A drinking game ensued in which the contestants sat at a table with a bottle of plonk and a bunch of glasses. At the drop of the flag the first glass was downed, followed by a race round some chairs at the far end of the bar and back to the table, where the winner was the first to finish the next glass. The glasses seemed to fill up all by themselves between sittings, and 'after a while, the floor was absolutely awash with white wine and everyone was quite plastered'.

Wyer went to sit down for his turn and someone pranked him by whisking the chair away. The great man pancaked onto the floor with a splash, and all the oxygen suddenly left the room. He rose to his feet in silence, eyeballing the crowd with his blood-curdling death ray stare. Then he cracked a smile and told them to get on with it. Resistance was futile. Any team member who attempted a night's kip was hunted down, dragged out and stripped. The party went on until everyone was fully basted.

The pounding heads packed up their troubles and returned across the Channel, where Wyer's headache was extended by having to deal with a labour dispute in the factory. This was doubly challenging since the company had never turned a profit. After a protracted delay, production got going again, and Astons had something to cheer: a new model.

The DB4 was their most expensive model yet, and sales immediately exceeded supply. In styling almost indistinguishable from the subsequent DB5, it had the best of Italian couture and British engineering. Aston Martin's distributors in San Francisco asked Wyer to ship them 1000 units, but were told that the factory could only produce 500 per year.

Wyer was asked for a strapline for marketing purposes and came up with, 'It will accelerate from a standing start to 100 mph and stop again in under 30 seconds.' Astons' press officer Alan Dakers had the figures published in the newspapers, so Wyer thought he had better prove it. He trotted off to MIRA, where the DB4 managed it in 27 seconds, proving Tadek Marek's engine and the adoption of disc brakes from the racing programme were all they were cracked up to be.

With a new model hitting the road and a veritable A-team leading the charge on the track, David Brown was not about to give up on his bid to beat Ferrari now. The main problem with the DBR1 was getting it to finish, and Wyer decided, 'On the premise that Le Mans is usually won by an obsolete car, we would not attempt to improve the performance of the DBR1 . . . [but] concentrate all our efforts upon reliability.'

That was all well and good, but over at Maranello the engineers would not be sleeping.

Winter Storm

The 1958 season closed out under a dark cloud for Ferrari that would lead to a big shake-up. The debt-ridden Musso had been driving every lap like it was his last. In Formula 1 he won the Grand Prix at Monza, nearly choking to death from a leaking exhaust in order to beat his English rivals. The next F1 race was Reims. The Ferrari mechanics had chosen sides, and a thinly concealed feud soured the garage. Romolo Tavoni, Enzo's team manager, merrily stoked the flames. On the morning of the race, Musso received an anonymous telegram: 'Drive at maximum.'

Hawthorn led from the start with Musso in pursuit and line astern to benefit from his slipstream down the long pit straight. At the end was a fast right-hander that only the bravest drivers attempted flat out without reducing speed. Hawthorn swept through. Musso had to adjust his line for a slower car but felt it was now or never to make his move. Instead of lifting, he kept his foot planted on the throttle. His Ferrari couldn't hold. It spiralled off the circuit and flipped over a ditch at 150 mph.

Hawthorn won and collected 10 million French francs in prize money to share with his co-driver. Musso was rushed to hospital, where he succumbed to his injuries. On her way back from the hospital, Musso's girlfriend Fiamma Breschi noticed Hawthorn and Collins having a laugh and playing football with an empty beer can outside the hotel. A month later Peter Collins was pursuing Tony Brooks at the Nürburgring Grand Prix when he lost control of his car across the hump at Pflanzgarten and crashed heavily. With his supreme reflexes Peter made very few mistakes, but this time it got away from him and he too was killed.

The 1958 Formula 1 Championship went down to the wire between Moss in the Vanwall and Hawthorn in the Ferrari. When Mike was disqualified at Oporto for driving in the wrong direction after a spin, Stirling stuck up for him and as a result his main rival retained his seven points. While today's F1 points system would have awarded Moss the title for winning four races to Mike's one, in 1958 the championship went to Hawthorn by one point. Mike immediately retired from racing but had not got over the loss of his close friend Peter Collins, and tragically, within months, he was killed in a road accident.

Enzo held a special respect for Hawthorn. When Mike high-lighted a fault in the Ferrari gearbox, Enzo told him he would never drive it again. 'Goodbye,' shouted Mike and walked straight out of the garage, knowing full well that Enzo would have him back. And he did.

Hawthorn's other complaint had been Ferrari's drum brakes, which faded out during races. His legacy at Ferrari was Enzo's adoption of Dunlop disc brakes for the 1959 season. This negated the advantage hitherto enjoyed by the Astons under braking. A knock-on effect was that Ferrari experimented with Dunlop's superior racing tyres and switched to these as well. These upgrades alone should have been enough to win, but Ferrari also set about re-engineering the Testa Rossa.

The V12 was given bigger choke carburettors to produce 340 bhp, compared to the DBR1s' 240, and the motor was offset left to accommodate a five-speed gearbox. The De Dion rear axle was now controlled by springs with adjustable Koni dampers, and part of the chassis employed space-frame tubing to reduce weight and increase rigidity. The exterior was remod-elled by Pininfarina to give it a more classic look with flowing lines from nose to tail, and more modest air inlets. Taken as a whole, the TR59 was a quantum leap forward in performance that should have made everyone in Feltham very worried. But they had no idea . . .

Annus Mirabilis

The minds at Astons were restless again. As 1959 dawned, David Brown was itching to get his new toy on the track and dusted off the F1 project known as DBR4. Its front-mounted engine design was out of step with the rest of Formula 1, who were already switching theirs to the rear, and it was underpowered by a modified version of Tadek's DB4 motor. Tadek had opposed its use in competition, but nobody listened. Reg Parnell liked the DBR4, which should have been a warning since he always liked the weird ones.

Reg was taking to semi-retirement like a bull in a china shop, and pig farming wasn't cutting it. He lobbied Wyer to let him take the DBR1 to race at Sebring. They went, it was hot, the car broke and Shelby walked back to the pits with the gear lever in his hand . . . The only plus point from the debacle was the wake-up call to David Brown Engineering, which finally widened the input gears to reduce the likelihood of shearing.

A chastened Astons team took its cars to the Le Mans test session in April 1959, with two months to go until the event. If they thought they could pick up where things had left off the previous year, they were sorely mistaken. Enzo had signed some exceptional speed merchants. He poached Brooks for his F1 team and kept his previous Le Mans winners Phil Hill and Gendebien together. His most dangerous pairing was Dan Gurney and Jean Behra. Gurney would go on to be the fastest driver in the mighty Ford camp at the Le Mans 1966 race. Jean Behra was an unstoppable force and, when he was in the mood, unbeatable. While racing on both four wheels and two, he had broken every bone in his body at least once and lost an ear. He had a prosthetic one, which he used for practical jokes.

At the Targa Florio in Sicily he was in full flow on his way down the mountain when he lost it through a fast left. The Testa Rossa slewed wide, flipped over a ditch and landed hard upside down in a field. Ferrari were preparing a rescue party when Behra reappeared in his battered car, assuring Brooks with the earnest guarantee of a second-hand dealer that it really was OK to continue. Brooks believed him, hit the track at speed and crashed when he couldn't steer. If Behra was highly strung *before* he joined Scuderia Ferrari, Enzo ensured he reached peak anxiety by refusing to confirm his number-one status in the team.

The Scuderia unleashed its full force at the Le Mans test day and lapped a crushing eighteen seconds faster than Aston Martin, leaving them plenty to ruminate upon. The DBR1 was slow on the straights, and there was nothing doing with the engine. That just left Enzo's famous put-down: 'Aerodynamics is for people who can't build engines.'

Ted Cutting went back to the drawing board. A plastic cover was fitted over the passenger seat, the windscreen was extended along the doors, and cowling 'spats' were fitted to the rear wheels for a cleaner air flow. A new doctrine, invented by a German aerodynamics engineer called Professor Kamm, specified extending a natural line from the tail of the car, then chopping it off at a right angle. This created a low-pressure area that produced less drag and more downforce. Wyer had some scale models made and they repaired to the wind tunnel. There was only one problem. David

Brown took one look at the truncated rear and said, 'I don't like *that*.' The line of the tail was raised but with a smoother finish than the full monty that Ted wanted. However, it would still be worth 400 rpm or 9 mph down the Mulsanne Straight at Le Mans, and that might make all the difference.

Moss had unfinished business at the Nürburgring, and since Astons were not going officially, he asked Wyer if he wouldn't mind lending him the spare DBR1 if he paid his own way. John went one better and sent him some mechanics and spares and went along in support, with Reg running the operation through his loudhailer. Stirling selected Jack Fairman as his co-driver, not for his speed but because he was 'steady' and on condition that he allow Stirling to drive as many laps as the rules allowed. For Moss, this was a personal test of endurance par excellence.

The Ferraris took some dialling in, their new shock absorbers providing them more lift-off than a jumbo jet. During qualifying Behra gave it both barrels to post a record-smashing time of 9 minutes 37 seconds. Moss only produced 9' 43", and most of the Scuderia, Brooks included, were quicker than that. Moss remained cool.

At the start of the race proper, Stirling sprinted across the track and left so quickly that the rest of the grid looked like a car park. He then proceeded to obliterate his own lap record from the previous year, caressing the brakes, maintaining momentum and carving his way down to a meteoric 9' 32". He put an outrageous lead of nearly six minutes on the Ferrari before handing over to his team-mate Jack, 'who would drive sensibly, and that was all I needed'.

Rain had started to fall, and Jack found himself in a spot of bother. As he searched for traces of grip on the notoriously challenging circuit and tried to get into a groove, the Ferrari drivers were tearing strips off his lead. Reg eyed the stopwatches intently, and then Fairman failed to come past.

Some considerable time passed. *Well, that's it*, thought Moss. He packed his helmet and gloves away while he counted the cost of his failed venture – wages, fuel bills, tyres, travel, insurance – and began to unzip his overalls.

'*He's coming in!*'

Stirling yanked his lid and gloves back on. Fairman hove to, and before the car had even stopped rolling, Moss dragged him out and was gone. Reg noticed a large dent in the DBR1's boot.

Jack was left to explain his detour to the pit crew. He had dropped down into the right-hander at Brünnchen, where he was blocked by a backmarker and ended up sliding sideways across the slick surface in slow motion, coming to rest halfway into a ditch. The DBR1 was beached, with its front dangling in the air and the rear tyres spinning helplessly. With his 'steady' reputation in tatters, Jack had tried to lever the car out with a fence post but it refused to budge. An act of faith was needed. Jack put his back into the boot and with his feet on the grass bank did a thigh press the likes of which Joe Wicks has never seen. The bastard came free, and off he went.

Stirling was frightfully good-natured about it afterwards, to the press at least. 'Here was a great chance for me to have a go at the sort of motor racing I enjoyed most – one Aston Martin against the full Ferrari team.' Privately to his secretary Valerie, it was a different story: 'You simply would not believe it, but that bloody idiot went off the road and stuffed the car down the banking. Lucky it was Jack, though . . . he's got the strength of an ox.'

The ox had left Stirling two minutes adrift of the Ferraris. He reeled them in, got past and restored a lead of nearly three minutes before handing the car back to Jack for his final two-lap stint. Phil Hill, a future Formula 1 champion, was going like the clappers in a Ferrari and, having passed Brooks, retook the lead from Jack . . . Moss got back in, and now the crowd were on their feet, 230,000 fans cheering him on, waving from every part of the track. He returned the compliment, and a mighty roar bounded along the grandstands as he passed. Nobody had ever seen anything like it.

Reg signalled the gap to Hill on Stirling's pit board, smiled to himself and shook his head in disbelief. The gap shrank by seven seconds a lap. Fellow driver Innes Ireland watched Moss approaching Flugplatz at 130 mph. 'As he came over the hump the car got terribly light, if not airborne, then it banged into the dip and up into the right, Stirling driving right on the door handles and with bags of opposite lock.'

Moss caught Hill there, hovered on his tail for a bit and sneaked through. Then he drove away and won by forty-one seconds, crushing the world's best drivers in what may not have been the fastest car but was the most throwable. There was no sign of physical fatigue in Stirling that year. He had evolved, or as the journalist Denise McCluggage said to Phil Hill, 'Don't feel too bad; you were the first human being to finish.'

False Teeth

I believe that if a man wanted to walk on water and was prepared to give up everything else in life, he could do that.

Stirling Moss

This would be David Brown's tenth attempt at Le Mans, and he prayed 1959 would be the breakthrough year. The team shipped across the cars, laden with gadgets, and arrived at their familiar second home at La Chartre. The operation was military in style, Reg's secretary Gillian Harris planning arrivals by the minute and ensuring everyone was in the right place at the right time.

Le Mans involved certain protocols. For Reg this meant dealing with French-speaking officials via an interpreter on a dusty parade ground while they poked the cars with blocks of wood to measure clearances, and shone lights into cylinder heads before sealing them. The Ferraris and Astons were closely matched at the weigh-in, both just under 1900 lb. Jean Behra prowled the arena like a bull, the French hero mobbed at every turn by adoring fans. After passing 'scrutineering', Reg got his life and his cars back.

At practice the Ferraris rolled out looking majestic. Behra's car was late for the test session and he was hovering around it like a cat on a hot tin roof. He was incensed by Tavoni's decision to run the same suspension settings on the cars that they had tuned specifically for the Nürburgring at a track like Le Mans, which had a totally different profile. And the cars were undergeared, running out of revs on the straight and costing time. He also had to fight to drive alongside Dan Gurney, knowing Gurney to be the fastest member of the team besides himself, instead of Cliff Allison.

When Cliff posted a faster lap time than he had in practice, Behra became suspicious. He organized a drag race between the two Ferraris down the Mulsanne Straight to compare engines. Cliff's was better. Behra drove to the pits, found Tavoni and punched him square in the face. Despite these ructions, speed was with Ferrari and their lap time of 4 minutes 3 seconds was considerably faster than the 4' 10" set by Moss, with Salvadori a couple of seconds adrift.

Back at La Chartre, Reg turned his attention to the game plan.

REG: When you see the arrow for fuel, you come in. Thirty-eight laps for Stirling, if you want to stretch it a little, we'll signal the laps.

FAIRMAN: How many laps on reserve?

REG: Two laps, maximum. Signals: hand on helmet if you go into reserve, point if you want to come in.

REG: Frère, Maurice, thirty-two laps per stint you two, *trente-deux*.

MOSS: Maximum revs?

REG: Six thousand. I don't mind six one hundred, but I'd prefer to stick to six for safety.

MOSS: But if we get a tow you don't mind six two?

REG: No. Ferrari will probably change brakes as often as we do. Save the brakes if you can, but don't tell me, 'I saved the brakes and I'm ten seconds a lap slower, wasn't I marvellous?' [Stirling had no intention of saving brakes; they were a resource.] This race has never been won at a greater lap average than 4 minutes 24 seconds. If you average 4' 20", I reckon you've won it. Finally, if any of you is not feeling well at four o'clock tomorrow, say so.

There wasn't a cloud in the sky so the wives unpacked picnics and stretched out while the drivers bathed in the river rushing past the hotel. Stirling appeared and dived from the bank into a pool of shaded water. It was icy, and as he screamed from the shock his front dentures fell out, the originals having departed in a previous racing incident. Stirling rang Valerie, his PA, and asked her to fly out the spares pronto.

Carroll Shelby was determined not to get dysentery for the second year running but his brown leathery complexion was growing paler by the second. He yielded to the famous Collins cubicle and switched to a diet of Coca-Cola for the remainder of the weekend.

Moss planned on playing Scrabble and turning in early. But before he did he took a last look at the DBR1 with its long, green, flawless features glowing in the evening sun. He went through the motions in his mind: running to the car, leaping the door, left leg straight past the wheel into the clutch pedal in one motion, crank and go.

The next morning Valerie arrived with Stirling's special cargo. 'Right, let's have them, Viper,' he said.

Val couldn't resist gently teasing the control freak. 'Errr, what?'

'My bloody teeth you fool! What the hell do you think you've flown out here for.'

She carried on the pretence for a while but eventually let him off the hook.

The team was ready to roll. David Brown led the convoy of pretty cars out of the village, his merry men wearing colourful bobble hats. They sped down the boulevard of tall poplar trees towards the track. The adrenaline was starting to flow. Crowds filled the grandstands and the two-storey buildings above the pit lane too. The drivers put on their thin blue racing suits and stared into the distance, into the future. The superstitious made their preparations, Moss secreting a little horseshoe inside the cockpit of car number four. Reg dipped the fuel tanks one last time, just to be sure. The gendarmes formed a line and cleared the gawkers from the pits.

As the clock inched closer to 4 p.m., the fans were teeming in the start-line tribunes. The drivers walked to their little white-painted marks and turned to face the long line of fifty-four pristine cars. Time slowed. Heartbeats hardened. National flags clinked against their masts. Each man was looking for a small advantage on the opening lap while knowing that a race can't be won in the first corner, but it could be lost there.

Silence as the starter flag rose. And dropped. Pitter-patter went the feet, Stirling was first in, gunned the engine clean and was away first. Behra's engine stalled. The mob were away and kicking up dust, flying over the hump below Dunlop Bridge before rushing down to the Esses, diving left, then right.

Stirling kept it clean, the full tank giving the DBR1 a 50/50 weight balance and understeer on cold tyres. He dropped to second gear into Tertre Rouge in order to get the nose in, minding the sand trap to the left, and opened the throttle on to Mulsanne.

Behra had got going and was charging through the field, slicing past cars indiscriminately at every curve. At the end of lap one Moss blew past the pits with the Ferraris of Ramos and Gendebien in hot pursuit, Behra already up to an unbelievable sixteenth.

By the tenth lap Behra was hammering along the Mulsanne Straight in the slipstream of the other two Testa Rossas. The rev limit set by Tavoni was 7500, and Behra thought him a fool not to have installed longer gears. He wasn't for easing off, the needle twitching as far as 9000 rpm, 188 mph, as he set a lap record of 4 minutes 3 seconds on his way to second place.

Moss looked left as he passed the Hippodrome café on Mulsanne at 166 mph and eased his foot off the accelerator to hold 6000 revs. In a few hours that bar would be swarming with brazen revellers saluting the cars with steins of ice-cold beer. He lined up for the tight right-hander, dug into the anchors at the 300-metre board, shifted the awkward gear lever down to first and steered the car in on the brake. Mind the exit.

One of the GT cars lay beached atop the sand dune on the outside. The French marshals handed the driver a shovel and curiously watched him dig himself slightly deeper with every stroke.

Reg looked down the pit straight for signs of his man emerging from White House Corner. Moss appeared with Behra's TR59 right up his tail pipe. Behra swerved out, and the French fans roared their approval as he passed the Briton. Behra swept into the fast right-hander under Dunlop with Moss now tucked in close behind, his right wheel shaving the white line at 150 mph.

Behra edged away over the crest, then Moss closed a little under braking for the Esses. The Ferrari looked awkward, and its tail hopped as it entered the left and clumsily changed direction, while Moss pitched through in a snaking, balletic four-wheel drift. He would have to drive perfectly now to ride on Behra's coat tails. The Aston's short gear ratios worked superbly, squatting the rear of the car down in the fast curves, giving him the grip he needed.

Behra slowed for Tetre Rouge. Moss felt the fuel tank lightening in the rear of the car, providing a more steerable weight distribution of 52/48. He closed up, taking the fast right in third gear rather than second. If he could just hold onto Behra's slipstream, he could stay with him.

Behra desperately needed to break the tow he was giving Stirling and kept his V12 screaming for mercy. Moss clocked his tacho: his speed was so much faster on Mulsanne thanks to Behra's slipstream this time round that he backed off before Hippodrome. All of Ted Cutting's aero modifications came into play, keeping him in the chase. The car felt perfect, but bloody hot around his feet.

Shelby felt like death and was drinking cola like it was a new sport. He had noticed that the car's exhaust system had been rerouted around the driver's feet and the cooling duct removed. As he pounded the laps he was sure glad he'd wrapped his racing shoes with asbestos, otherwise his feet might have been blistering as badly as those of one of his team-mates, Maurice Trintignant. 'Trint' was in agony. After whingeing to Reg, expecting sympathy, he was nearly replaced but had chosen to soldier on.

Two hours into the race as the afternoon faded, Behra pulled the pin out of his grenade and set the fastest lap of the race at 4' 0.9". How his patience was tested, however, when Ferrari called all their cars into the pits at the same time . . . Wheels and jacks went in all directions, accompanied by a multilingual chorus of profanity. Someone checked Behra's tacho: 9300 rpm. Gurney took over from Behra.

Jack 'Ox' Fairman replaced Moss and lost a little ground to the rapid Gurney, kept ahead of Jaguar, and was followed by Shelby in fourth. Moss climbed back in and did what he did best, tried to close the gap by driving flat out.

At 10 p.m. Moss accelerated through to top gear, and the DBR1 shuddered along Mulsanne, the little headlights capturing the strobe effect of the white dotted lined blurring underneath. Suddenly the air felt heavier, like he was straining against an elastic band. He checked the round gauges in the spartan dashboard, confirming what his heart already knew. Down on revs, oil pressure off, engine valve on the way out.

You must know by now that Moss was different to other drivers, and Ferrari knew it too. For a misdirection to succeed against your opponent, your story needs to be convincing so that he doesn't notice he's being distracted. Moss had been given a special low-friction four-bearing engine, rather than the standard seven, that could run a higher compression ratio for a modest improvement of 15 bhp. In short, Wyer had used Moss as an irresistible hare to lure the hounds, who were now frying their engines. Ferrari were on borrowed time. Moss was 'expendable', a sacrifice for the greater good of the team, but spare your pity because he wouldn't have wanted it. 'I would rather lose a race driving fast enough to win it than win a race driving slowly,' he once said.

Stirling got back to the pits, his race run. Wyer was hunched over an array of stopwatches and time sheets. After a brief explanation, Moss asked how it was going. The lead Testa Rossa blared past the pits emitting its heavenly howl.

'Alarmingly healthy,' Wyer replied caustically.

Moss got changed and spent the night loitering around the Ferrari pits, pointing and shaking his head at imaginary faults, taking an occasional break to visit the trackside fair and strip-tease show. Gurney pulled slowly into the Ferrari pit, waving for someone to help him, but the pit was empty so he drove off again.

One of the Aston mechanics pointed this out to Wyer.

'Nothing trivial, I hope,' he replied.

Gurney returned, was given a metal tube to replace his gear-stick and motored into the night. Shelby meanwhile was hitting rock bottom. As he sped towards Indianapolis he felt the pain of a knife stabbing slowly into his chest, reached into his pocket and swallowed another nitroglycerine pill. *Goddamned angina.*

Behra took over the lead Ferrari and drove like hell until about 1.30 a.m., when he blew a gasket. He came into the pits with steam pouring out of the exhausts. He was done. Phil Hill's Ferrari led the Aston of Roy Salvadori, who had taken over from Shel.

Roy heard a tapping going down Mulsanne. He switched off his engine, but the tapping continued ... then it stopped, and came back. He pitted to check the transmission.

'Get back out there and come in at the proper time,' Reg told him.

'But—' Roy protested.

'*Off you go!*'

The tapping got worse, and the car was slithering around in

the corners, the vibration becoming so bad that he couldn't see the track. Roy suspected the suspension was broken.

When Roy came in at the end of his stint, Reg had the car lifted up on jacks, got in and gunned the engine to spin the wheels. This caught the attention of the obese French *commissaire* assigned to police Astons' pit procedures. The *commissaire* might have had more to say about what Reg was doing if the boys hadn't plied him with enough ouzo to stun an elephant. As it was, he just leaned against the wall with his cap drooping over one eye and grinned. The crew spotted some balled-up rubber that had broken away from the tyre . . . changed it, and sent Shelby out.

Maybe Roy should have known better, but there was no time for a discussion. Phil Hill and Gendebien had driven more cautiously than Behra and now had a clear lead, which they kept through the night and into the morning.

The waft of BBQ smoke drifted across the track as bleary-eyed fans crawled from their tents. Then, at 11 a.m., the gauges on Hill's Ferrari started flashing red and he came in. The bonnet went up. The engine was deep-fried. The Jaguar was out too. *Aston Martin were leading Le Mans – one, two.*

The British fans began their unique, raucous, tribal cheering. Shelby took over from Roy in the lead, tracked by Aston teammate Trintingant, who was given the E Z E board, telling him to hold formation and preserve the car. But Trint sped up. Shelby responded. Reg considered throwing his loudhailer at the Frenchman; instead he called the signalling station and had some salty messages translated into French that seemed to take effect.

The Astons cruised round at a set pace, and at one minute past four on the afternoon of Sunday 21 June 1959 Carroll Shelby crossed the finishing line of the Le Mans 24 Hours in first place, followed by Paul Frère in the second Aston. Ten years of work had paid off, and so had the wily plan to deceive the Ferraris devised by Wyer and executed by Reg.

In the euphoria of victory Roy Salvadori went missing, while David Brown was offered a victory ride in the winning DBR1 with Miss Europe on his lap but sat down in a puddle of oil, ruining both his tweed suit and the moment. Shelby took a few giant swigs of champagne, ran around 'drunker than a billygoat' and fell asleep on the trackside in just his underpants and leather shoes. 'It didn't mean much to me at the time. I hardly thought about it.' The time would come when he thought about little else.

Reg climbed into the team's Lagonda to leave the circuit, turned to his secretary Gillian to say something deep and meaningful, and passed out. There was no party at Le Cheval Blanc. Everyone was too zonked.

David Brown felt a mix of relief and joy at having taken Aston Martin to the pinnacle of motorsport. There's no more enthusiastic racing crowd than at Le Mans, and as the Astons returned to La Chartre they were treated to the warmest reception by the great people of the Pays de la Loire. It was tradition for David Brown to drive the lead car away from a race, and he is said to have found its condition so good that it could have done another 24.

At 5.15 p.m., just one hour after winning the biggest racing event of his life, D. B. took off in his private plane, and as the pilot banked over the circuit he turned to Wyer and said, 'Thank God we need never go to that bloody place again.'

PART SIX
1960–88

Bulldog

THE AGE OF GIANTS: FORD AND THE RISE OF THE MUSCLE CAR

Aftermath

It was as if a balloon had been expanding for ten years, and then suddenly popped. The 'whatever it takes' approach had consumed every spare hour of Wyer's time, seven days a week, for years. Every favour had been asked, every coin expended. They had expected to win in 1958, and this time round the sense of relief was palpable.

Wyer and Reg kept immaculate records of the races, and in my research for this book I only found one typo, where Reg crossed out the word 'difficult' to emphasize an important point: 'It is *impossible* to over-estimate the part played by Moss in our success at Le Mans. He set a very high average speed in an attempt to break up the Ferrari opposition, and his performance was without doubt responsible for their failure.'

They looked inside Stirling's engine and found a tiny shard

had broken off the rim of an inlet valve and done the damage. Salvadori believed, 'Moss was very unlucky. He was very gentle on his car and did not push it unduly,' having maintained the correct revs throughout the race. Shelby was more sanguine: 'Le Mans is 80 per cent luck. A car that's built for racing is not supposed to stay together for 24 hours.'

But Shelby's did, and it was plastered across newspapers around the world. Aston Martin also found themselves leading the World Sportscar Championship thanks to their accidental entry and win at the Nürburgring, courtesy of Moss. With one more round remaining at Goodwood, they decided to shoot for the title.

The Goodwood circuit is a 2.4-mile sequence of fast corners without much of a straight to favour the Ferrari V12. The constant curves and heavy braking would suit the Astons as long as they didn't catch fire, which had become a Goodwood ritual. Moss and Salvadori were paired to create the fastest line-up in car one, while Shelby joined Fairman in car two and the French lads went in car three.

Reg had been watching the Indy 500 on the telly and been impressed by the pit work. The American crews plugged a compressed-air line into their cars to power four hydraulic posts that extended from the floor and lifted it off the ground. Goodwood had an abrasive surface so using these would really speed up the tyre changes. For reasons unknown, Reg also recruited some delicate hands from the factory to work the tyres, hot brakes and the fuel rig.

According to Wyer, 'Shell asked us to go to Goodwood and test some new refuelling apparatus, thinking, I suppose, that if we could refuel a car without setting fire to it anybody could.' The new system had a safety release to control the flow but could deliver 40 gallons in just 11 seconds. Having practised with the dummy-proof version, Astons pitched up for the race and discovered they had to use an older rig without a safety valve.

It was a sunny Sunday afternoon in Sussex, and the main protagonists from Le Mans had descended on Goodwood for the comparatively short six-hour race, except for Behra who had proved too feisty even for Enzo and been elbowed. The race was due to start at the stroke of noon, and the punctilious Moss synchronized his watch with the control tower so that he was ready to start running on time, anticipating that the dignitary who had been invited to throw the starting flag might not be quite so conscientious. Sure enough, as the starting flag waved, the patter of one man's feet had already kicked off proceedings. The other drivers, humbugged yet again and completely by the book, followed in his wake.

Roy took over and maintained Stirling's commanding lead until the time came for the pit stop and driver change. Until that weekend Bryan Clayton had never held anything heavier than a pencil in the comfort of his draughtsman's chair. Now he staggered towards the DBR1 with a heavy fuel hose under one arm and a dipstick in the other, while the crew rushed in to hammer the wheels loose. The searing hot exhaust pipe below the driver's door belched a fiery burp as Roy cut the engine. Bryan was oblivious to the fact that his lever was turned on and the fuel flowing freely.

'The first thing I saw as I was fumbling with the filler cap was a flicker of flame running down the back of the car. I thought,

That shouldn't be there,' then *whoosh*, up she went. Stunned by the initial blast, the crew then realized what was happening and fled, none faster than Roy, who was fully lit. He leaped over the bonnet, kicking his mechanic in the head, and rolled on the grass to extinguish the flames. The car was ruined and the garage torched to the ground. Was it game over? Come on, this was Aston Martin.

In a hugely generous gesture, Graham Whitehead immediately withdrew his privately entered DBR1 from the race and handed his pit over to the works team. Reg still had two cars in the mix. Fairman was due in with the second.

'*Moss!*'

Stirling looked over, hungry and leaner than a butcher's dog. Reg pumped his fists up and down beside his head, code for 'Get in the car right now.'

Moss rejoined the race aboard Fairman's car in fourth place and treated the audience to a repeat of his Nürburgring demo. He lapped the course two seconds quicker than anyone for the next three hours, taking the lead, the chequered flag and with it the title. Tavoni had to call Enzo again, and a minute's silence was observed throughout Maranello. After an epic ten-year journey under David Brown's leadership, Aston Martin were World Sportscar Champions.

The DBR1 was the finest machine that Feltham had ever made. Driving a lap around the Ring seemed easy by comparison to manhandling the brawny Mercedes 300 SLR or the more powerful Testa Rossa. It was an intuitive driver's car that inspired confidence, with fantastic brakes, balanced cornering and a strong connection through the divining rod of the steering wheel. Only five were made, which may explain the recent sale price of chassis number 1 for $22,550,000, making it the most expensive Aston ever sold. The ugly duckling Atom that David Brown bought in 1947 had spread its wings.

The hero of the hour was undoubtedly Stirling Moss. A household name, a four-times runner-up in the F1 Championship, and in the twenty-first century the voice of kids' show *Roary the Racing Car* and the old gentleman you bumped into around Goodwood. He won over half of the races he finished and redefined the limits of motor racing in an era brimming with talent. Moss was way ahead of his time, if not the greatest of all time.

Tony Brooks: 'The real essence of Stirling's approach, and where it was so different from mine, was that he was happy to live in that little area right on the outside of the envelope and I never was.'

Fangio: 'You measured your greater or your lesser capacity according to how you showed up in a contest with Moss – he was the driver's yardstick.'

Wyer: 'As a driver, in my opinion he had no equal at all. He was more than just a superlative driver, he was a great tactician, strategist and a race-winning factor.'

Enzo even offered to paint his car blue if that's what it would take to get Moss into a Ferrari. It never happened. Sir Stirling Moss, as he subsequently became, had a massive shunt caused by a mechanical failure in 1961 that ended his driving career.

Expensive Hobby

One must always anticipate what is coming up in front,
whether it is a corner or other cars.
David Brown

Motor racing was costing D. B. nearly £2 million a year and his decision to withdraw from sportscar racing after winning the world championship would have made financial sense, had he not rekindled his Formula 1 ambitions. Following the DBR4's failure, the team developed an even worse single-seat race car in the DBR5. With its uncontrolled independent suspension, its wheels pointed in all directions bar the one intended. Jimmy Clark had a close shave by signing to drive it, but even Reg didn't have the heart to hold him to his contract when Lotus, for whom he would win two F1 titles, asked Jimmy to drive for them instead. The Formula 1 project was canned in 1960.

The name Aston Martin nonetheless stood for glamour and success as never before. Over 1000 DB4s were built, including more aggressive models like the DB4GT, and the GT Zagato, which was styled by a coachbuilder in Milan. Only nineteen of the latter were sold, each for the premium price of £5,470. One today will set you back around £11.5 million. The GTs weighed 85 kilos less than the standard DB4, and with enhanced triple-Weber carbs could accelerate from 0 to 60 mph in just 6 seconds, performance that would not be beaten by another Aston until the 70s.

The overhead costs of top-flight engineers and skilled labour combined with low-volume sales and minimal profits were causing financial problems. The decision was made to increase production, and since the Feltham factory had reached its maximum output of twelve cars a week, D. B. moved the business to Newport Pagnell, where he had bought an auto engineering firm called Tickford. The problem was that increasing volume didn't reduce costs because everything was still hand-made. Until Aston Martin solved the riddle of producing high-end sports cars that turned a profit in a fickle market, it would remain an expensive hobby for D. B.

V8

Manufacture of the DB4 had been delayed by labour issues since its initial launch at the London Motor Show in 1958. It finally came off the line in 1961 in a bit of a hurry with a defect in the engine bearings that only materialized some months afterwards. Customers rang in so frequently that Wyer winced every time he heard the phone. Tadek Marek was dispatched to investigate and cheerfully called Wyer from Paris with an explanation: 'They all happened on Good Friday!'

Owners were enjoying their cars so much they were driving them fast over long distances, notably on bank holidays. This revealed an issue that had been masked during development. Tadek fixed it easily by regulating the oil temperature. The DB4 engine had never been his main focus, and he had always anticipated growing it into the 4-litre that powered the DB5. By the time the DB5 exploded onto cinema screens in James Bond's hands in 1964, Tadek was already thinking three steps ahead. To him the DB5 now felt a bit tame, and his solution was to hot-rod it by fitting a V8 engine inside an experimental prototype that became *his* DB5.

My default response to any film producer wanting more energy in a car chase is, 'Put a V8 in it.' The two banks of four cylinders are tilted 90 degrees apart to form a V-shape, and the benefit over a straight-six configuration is that the energy from the pumping pistons is harnessed at closer intervals, with smoother, stronger power delivery from a relatively small package. It can turn a lame duck into a soaring eagle and worked wonders in everything from the Daytona I drove in *Fast & Furious 6* to the 'speeder' in *Solo: A Star Wars Story*. Yes, a V8 went into outer space.

The *Star Wars* special effects team built a chassis that resembled a bed frame and placed a V8 in the front of it. The four-wheel drive system had a centre differential so that I could alter the drive from all four wheels to rear or front only with the flick of a switch and at full pelt in order to gyrate the speeder in all manner of unearthly directions. The body was styled by the art department, who doffed their caps to the muscle cars of the 60s with a deep-metallic-blue and pinstriped paint job. It was a stunning piece of engineering, and the rally-grade suspension meant that when I jumped it into an imperial TIE fighter factory (a lifetime ambition) the landing was as soft as a five-star mattress. The speeder's wheels were painted out of the movie using CGI, but all the action was real.

The 1960s was about fast cars, short skirts and rock 'n' roll.

David Brown saw what was coming and realized that Tadek's V8 engine needed a new home.

The 1966 DB6 was similar to the DB5, with the same motor and styling but with an aerodynamic Kamm tail and stiffened De Dion suspension. Astons' chief designer Harold Beach then used this chassis and suspension to develop a new model that could deploy Tadek's more powerful lump. He brought in a brilliant stylist called William Towns, at first to design the interiors and door handles, but his art soon found truer form. Towns broke away from the bloodline that ran through the DB4, 5 and 6 and proposed a butch fireeater.

Had logic prevailed it would have been called the DB7. However, Touring of Milan were originally instructed to style the car with their Superleggera lightweight body, so it acquired an S to become the DBS. Ironically the finished product was 200 kilos heavier than the models it succeeded, partly because the jigs for building the car were accidentally made three inches wider than Towns' drawings . . . Everything was still hand-made.

The DBS V8 that finally waded onto the streets in 1970 was a powerful brawler that matched the DB4 GT for pace, with 0 to 60 mph in 6 seconds and reaching 100 mph 8 seconds later. The torque going through the rear wheels bent the steel-wired rims, so it was refitted with sturdy alloy spokes instead. It made a shy appearance in *On Her Majesty's Secret Service*, and Q didn't invest much in gadgets, the lack of bulletproof windows proving fatal to Mrs Bond, RIP.

Tadek's 5.4-litre V8 engine would power *all* Aston Martins for the next *twenty years*, and during this time the pace of development of the DBS line was glacial. David Brown had been knighted in 1968 for his services to industry and was spending more time on his core business than on the car side. Fluctuations in sales were killing Astons' bottom line and in 1971 the business was put up for sale – again. Fortunately, D. B.'s timeless flair and the cachet of a certain spy had left an indelible impression, and there would always be buyers for both Aston Martin cars and the business.

At this crucial juncture, only a business with experience of mass-market production could rescue Aston Martin from extinction.

Ford

The Lincoln limousine peeled off the Edsel Ford Expressway, cruised along Ford Road and turned into the parking lot of Ford Global HQ in Dearborn, Michigan. Henry Ford II – 'Hank the Deuce' – swivelled his legs off the sofa-like back seat and marched towards the tall shining edifice aptly known as the Glass House. By 6 a.m he was sitting behind the desk in his office on the twelfth floor with his father's portrait on the wall looking over his shoulder. Surrounded by 2000 acres of industrial complex dedicated to the family business, he contemplated the situation.

Back in 1945 Ford had been haemorrhaging $9 million every month. Hank's father, Edsel, had tried to modernize the business but been prevented from doing so by Henry I, who was increasingly detached from the world, travelling it on his own private train. The Deuce was no engineer, but as a student of sociology and the son of gentle parents, he had qualities of empathy and leadership. When he took over the business, he called in the company enforcer Harry Bennett, who had bullied both his father and Ford's employees, and fired him.

'You're taking over a billion-dollar organization that you haven't contributed a thing to,' Bennett had ranted.

The man had a point, so the Deuce hired a management team of whizz kids, ten of the brightest analytical and logistical minds from the US Army Air Force during World War II. They researched market trends and customer data, enabling Ford to sell products that were relevant. While Henry I had aspired to Ford owning every element of its supply chain, right down to sourcing rubber from its own plantation in Brazil, the Deuce was a champion of outsourcing. But the clock was ticking. It was 1962. GM had won most of the NASCAR events that year; Ford had won zilch, and GM were selling twice as many road cars as he was.

Carroll Shelby had already conceived the idea of mating a powerful American V8 with a European sports car and approached John Wyer about using the DB3S. This was a tidy, if underpowered, little racing car that begged to be produced for the road. It's best not to dwell on what could have been, had this heavenly concept gone ahead. In what Wyer recalled as one of his worst ever decisions, he declined Shelby's offer in order to focus on rectifying the DB4.

Shelby went to another British firm, AC, which was all too happy to offer him its Cobras. Shelby then approached Ford, which in 1962 bought into his vision of an American GT car and agreed to supply him with its small-block V8. The result was a powerful little monster that would bite you if you were less than on full alert behind the wheel. The Shelby Cobra was low to the ground and as brash and noisy as its creator.

Ford adopted a dynamic slogan – 'Total performance' – to connect their motorsport programmes with the youth of tomorrow, and it was manna from heaven when the Shelby Cobra 'powered by Ford' outpaced and outraced GM's leading model, the Corvette.

The Deuce reputedly then saw a race during which an Italian sports car rocketed into the distance and said to one of his team, 'That's the way to go racing. Why don't we buy those red cars?' Ferrari was indeed for sale, and Ford sent a large delegation to Modena to discuss terms. The negotiations resulted in a final figure of $10 million, chump change for Ford, for a 90 per cent stake in Ferrari's auto-making business and 10 per cent of the racing arm. The Ferrari-Ford racing team would teleport Ford's business into the Mecca of motor racing at Le Mans.

Enzo focused on a detail in the contract: 'If I wanna go race, it is written I have to request authorization from America. If you do not wish me to, do we go or do we not go.'

The reply from Ford's Don Frey was, 'No, you do not go.'

According to Enzo's right-hand man, Franco Gozzi, Enzo's volcano blew in as spectacular a fashion as he had ever seen, before or since. Using words that 'cannot be found in any dictionary', Enzo lambasted the insult to his integrity as a manufacturer before turning to Gozzi and saying, 'Let's go eat.' Whether it was a genuine off-the-cuff reaction or a longer-term

'Fill 'er up.' Lionel Martin has more to say from the sideline during a pit stop at the French Grand Prix at Strasbourg, 1922, with Kate by his side, as ever.

Augustus Bertelli and Pat Driscoll won the Biennial Cup at Le Mans in 1932, with the smiling Sammy Davis overseeing proceedings as team manager. Spot the lashings of rope holding the bodywork together.

Big gear, little gear, DB6. David Brown Engineering built gears for all manner of applications; this is the gearing that rotates the top of the BT Tower in London.

Fearless and peerless, Count Louis Zborowski strikes a pose in Chitty 1. She must have been quite the handful to drive, and there was no missing the drain-pipe exhaust.

Smoke billows from the exhaust as Lou puts the pedal to the metal and slides his Grand Prix car up Shelsey Walsh in 1922.

An exhausted Carroll Shelby celebrates Le Mans victory in 1959 with his team-mate Roy Salvadori and the 'Patron' David Brown. Having abstained from food for over 24 hours, that champagne knocked old Shel pretty sideways.

Right: They thought it was all over… Jack 'The Ox' Fairman desperately heaves the DBR1 out of the ditch at the Nürburgring in 1959 before handing over to Moss, who duly performed a miracle by re-claiming the lead and taking victory.

Below: Kindred spirits: Juan Manuel Fangio congratulates Stirling Moss after winning the Italian Grand Prix. Between them they elevated Motor Racing into a professional sport and set the bar with their code of honour.

Left: Parkour! Reg Parnell (far left) looks on with trepidation as Pat Griffith puts a judo move on Peter Collins at the hotel during the 1953 Mille Miglia. John Wyer (far right) has seen it all before.

Below: Jackie Stewart provides feedback on the wild child 90s Vantage. Judging by Walter Hayes' expression (far right), it's bad news. Nick Fry keeps his arms folded in the background behind Adrian Reynard, who also assisted during development.

Darren Turner clobbers the kerbs at the Ford Chicane during Le Mans, on his way to victory in the GT1 class in 2007.

The author living the dream aboard the Vulcan at Thermal Club on one of those days you wish would never end.

The Moss method. Mastering the four-wheel drift was one thing, but slicing past the barriers – as seen here at the Esses section of Le Mans – and being pitch-perfect every time was quite another. Nobody did it better.

The running start for Le Mans 1958. No prizes for pointing out Stirling Moss.

Left: Mission: Shanghai. Bond takes delivery of an important package, courtesy of DHL.

Below: Q's 'winterised Volante' from *The Living Daylights*. The AMV8 was the perfect combination of brute muscle and suave sophistication. She cruises back onto the big screen in *No Time to Die*.

The Vantage was the T-Rex of the 1990s. You didn't drive the Vantage; it took you along for the ride.

tactic to squeeze a better bid out of rival bidder Fiat, the result was dramatic.

The deal collapsed and the Deuce gave Don Frey new marching orders: 'You go to Le Mans and beat his ass.' The trouble was, they didn't know how.

Ford v. Ferrari

Carroll Shelby introduced Wyer to a Ford executive nervously investigating how a company with no racing experience outside NASCAR might go about winning Le Mans. Wyer advised Ford to deploy a 'compact unit with a very high degree of autonomy, free from interference but able to draw on the resources of the parent'. They listened and then did the opposite.

During these conversations Wyer learned of Ford's abortive attempt to buy Ferrari. Aston Martin had just ceased all racing operations and was in financial difficulty. Wyer saw an opportunity to solve both problems by selling the business to Ford, but it all seemed too far-fetched to David Brown, and a second opportunity to align Aston Martin with Ford was missed.

Wyer's feet were going to sleep at Astons, so when he was asked to head up Ford's Special Vehicles Activity unit, based in Slough, he signed up. His mission was to create another Ferrari beater: the Ford GT40. After two disastrous years of racing by committee, Wyer's original suggestion of placing a small autonomous unit in charge of the campaign was heeded. Carroll's racing enterprise Shelby American took over in 1964, with Wyer reprising his role as head of European Operations for Shelby and keeping an eye on the development and production of the GT40.

Ford's Le Mans campaign against Ferrari was as close to a war as two companies could get and was immortalized in a recent movie by director James Mangold, so for a glimpse of the race that defined sportscar culture within Ford, we bounce over to Hollywood to the set of *Ford v. Ferrari*.

Rather than fly everyone to France to the real Le Mans circuit, the producers used a series of locations across America and blended them together in the edit. The pit lane scenes were filmed north of Hollywood at a film ranch where a lot of Westerns were made. If you headed up there early it was worth a detour via Topanga to follow in Steve McQueen's tyre treads up the winding canyon roads and warm up your skills.

The white picket fences, the grazing horses and an easier pace connected you with the Old West. In the middle of it all, the construction and art departments had knocked up a full-scale rebuild of the Le Mans pit lane with all the garages and open galleries decked out with team liveries and paraphernalia. Never mind the sun-bleached hills in the background; they could fix that in post-production.

I was playing the role of New Zealand F1 ace Denny Hulme, team-mate of one of the leading protagonists, and it was my turn to drive the pale-blue and orange Gulf-liveried Ford GT40 Mark II. From the outside it looked exactly like a real one. I slid across the riveted aluminium door panel and landed in the chair scoop, reached across and caught the exposed section of wire that was the door handle, pulled it closed and remembered to duck so that the roof part of the door didn't guillotine my temple. The door was a unique design feature – by opening a section of the roof it made driver changes faster.

The dashboard was a flimsy moulded-aluminium sheet to which the basic controls were attached by a mix of glue, some cable ties and a lot of hope. Some of the buttons worked some of the time. I flipped the silver toggle marked O N to on, and pressed the round black button next to it. The V8 behind me gurgled to life. I peered through the scratched Perspex windscreen and snapped the motor to ask Shelby's chief mechanic, a stuntman called Frosty, to move out of the way, which he did with a glare. I clunked the black ball gear lever into first and headed off down the pit lane.

I made a couple of left turns on the perimeter road around the dusty airfield and then aimed back towards the enormous and authentic-looking pit structure, teeming with mechanics, cars and 'fans' dangling over the edge of the galleries overlooking the pit lane. I ran my line onc more time for safety, having had weeks of voice coaching, then swung towards the Ford pit. The brakes screeched, I released the five-point harness and leaped out.

Carroll Shelby was overseeing proceedings in smartly pressed slacks with a shining buckle, wearing a Stetson and aviator sunglasses. He was a fraction shorter than I expected and bore an eerie resemblance to Jason Bourne. My team-mate was a rare bird. Ken Miles was a talented engineer trapped in the body of an exceptionally fast driver. He had been chopping and tuning sports cars since boyhood and had helped Shelby's Cobra eat Corvette's lunch in SCCA racing. He was drinking tea, as usual. His blue race suit hung from his gaunt, sinewy frame. He resembled a slimmed-down Batman.

The boys in dark blue Shelby overalls serviced the car with fuel and attacked the locking wheel hubs with lead hammers. I went straight to Phil Remington, Shelby's chief engineer, and delivered the immortal words 'The engine's getting hot, and the brakes are boiling' in my finest Kiwi accent and said a few encouraging words to Ken Miles, who was staring a thousand yards into the yonder.

'*Cut!*'

My star-struck observations of Shelby and Miles were in fact of Matt Damon and Christian Bale. Christian had done his homework on Ken Miles, mimicking his Midlands accent and quirky stoop. He was discussing a point with Matt Damon and leaned over to ask me, 'What would you feel towards another team when one of their cars has a big shunt out there on the circuit?'

I knew what John Wyer would say – 'Nothing trivial, I hope' – but that was just black humour. I explained that when something bad happened, racing teams closed ranks because you were all in it together. The humanity in that resonated with Christian and Matt and their interactions with 'Enzo' and the Ferrari team felt genuine. Both actors had dialled in to what racing was all about – obsession – and Christian had the racer's grit.

Playing the part of Ken Miles put Christian Bale behind the wheel for plenty of shots, and stunt coordinator Robert Nagle, of *Baby Driver* fame, ran him through his paces in a reasonably priced sports car: the lightweight and nimble Mazda MX-5 Miata. It was ideal before climbing into the more aggressive Cobra that he drove for the pack shots round Willow Springs. In typical Bale style he picked everything up quickly and was indiscernible from the rest of us. He also made some method-acting notes on the racing-driver breed by sneaking into our morning briefings and peering over our shoulders. I don't think it took him long to size up the twelve grown men pushing Hot Wheels toy cars around a whiteboard and haggling over their positions.

Beyond California, to double corners like the Esses and Tertre Rouge we used a bug-infested track at Savannah, Georgia dressed with sandbanks and straw bales to make it look the part. Nagle had created a biblical playbook of choreographed moves for us to memorize since everything had to be historically accurate, and assembled a crack squad of pro drivers to pick up these moves with minimum fuss. These included sons of legends from the original event like Alex Gurney, who played his father Dan, Derek Hill to play Phil and Jeff Bucknam as Ronnie – all were demon drivers in their own right. Alongside them were seasoned film regulars like Jeremy Fry, who nailed those pitch-perfect moves in the red Subaru on *Baby Driver*, as well as Le Mans winners, drag racers, rally champions, etc.

It was 90 per cent humidity and the primeval cars had no air con, so you had to decide whether to open the door and invite in a plague of love bugs to have sex on your sweating face, or leave the door to the sauna closed. I opted for the latter since Nagle worked fast. We would tear down the main straight and brake with varying levels of efficiency, float through a corner as a pack and shuffle for the cameras according to the playbook.

To represent the Mulsanne Straight, we moved to a big stretch of road in Statesboro, where our stock V8s propelled us up to 180 mph, not far off the 212 mph the original racers were achieving down the Mulsanne in 1966. Ours just rattled a lot more. The local sheriff helpfully explained that the main difference from France was that here 'If your car breaks down and you stray into someone's back yard, they might take a shot atcha.' I monitored the fuel gauge more closely after that.

Le Mans was a race of attrition, and knowing this, Ford had entered thirteen cars in the 1966 race. If this seemed like over-kill, it was vindicated by the result, which saw Ford place in first, second and third while none of their other entries finished. For Ken Miles it was a bitter-sweet end to a long campaign. Working tirelessly with Shelby's brilliant engineer Phil Remington during zillions of miles of testing on lonely airfields, they had ironed out the suspension failures, wayward handling, bodywork flying off and gearbox issues that had plagued the early prototypes.

Remington also invented a clever system for replacing worn brake rotors in one minute, a marked improvement from fifteen. Only then was the GT40 capable of sustaining 24 hours of abuse, with 9000 gear changes and 8 million piston revolutions to sus-tain an average winning speed of 125 mph.

The strategy Shelby's team used at Le Mans was the same win-ning formula he used in his Aston Martin days: send the hare to break the Ferraris. Ken's speed made him the hare, but unlike Moss, his car held together and gave him a four-lap lead by the closing stages of the race.

Ken was ordered to slow down for a photo finish of all three leading Fords and, according to Wyer, 'Bruce McLaren out-fumbled Ken on the line and won by 20 metres,' denying Miles a place in history as the first driver ever to win the big three in one season: Daytona, Sebring and Le Mans. Ken's career looked like it was on course, but just two months later he was killed during an accident in testing. Ford would win Le Mans four years in a row, embedding the GT chromosome into their DNA and placing Aston Martin firmly on their radar.

Operation Vantage

The 1970s was a challenging time for luxury car makers, with oil price inflation of 400 per cent, another stock market crash and a seismic shift in consumer demand towards smaller models. A more timid firm might have considered a change of product strategy but not Astons. In 1978 they doubled down by releasing Britain's first supercar: the V8 Vantage.

'Vantage' had previously denoted a performance variant of an existing Aston Martin model, but the V8 Vantage had a gravity all of its own and created an altogether new line that would evolve and eventually be reborn into the must-have icon that elevated my dad's shopping trip for socks into a life-changing experience.

The existing V8 engine was given a higher compression ratio, thirsty Weber carburettors, modified pistons with bigger inlet valves and stronger camshafts to raise the bar from a paltry 300 to a healthier 380 bhp, which the firm described at the time as 'adequate'. In case the car didn't look mean enough already, the rear wheels were widened. It was clocked reaching 60 mph in just 5.2 seconds, and with a top speed of 170 mph was faster than the

Lamborghini Countach and any of the latest Ferraris. William Towns nailed the V8 styling with a lusty bonnet scoop and curt rear wing neatly integrated into the bodywork.

It still wasn't enough for Towns. He foresaw the hypercar before the word existed. Rather than construct a concept car for show purposes only, he built a running prototype in 1980 with the first rear-mounted engine in Astons' history. The wedge-shaped, gull-winged futurama named the Bulldog was powered by the same 5.3-litre V8 motor as the Vantage but boosted by twin Garrett superchargers to achieve a mind-boggling 700 bhp and a theoretical top speed of 237 mph. Its closest relative would be another ten years in the making with the arrival of the McLaren F1.

Typical of Towns, specialized features were embedded into the angular bodywork such as a bank of five centrally mounted headlights that would have been visible from space. Sadly there were not the resources to bring it to market, and only one of these otherworldly machines was made. Its whereabouts remains a mystery but I'm led to believe that a mission might just lead to its recovery one day.

None of these exciting announcements changed the fact that Aston Martin was still in a slow-motion car crash, financially. It took three months to build one car; in fact managing director Alan Curtis took pride in 'never building more than seven a week'. Through the 70s and 80s sales ranged between 80 and 200 per year, which may have been good for wealthy collectors, but wasn't sustainable as a business.

Victor Gauntlett was a larger-than-life entrepreneur who came into Aston Martin during those troublesome days and kept the company going with sheer enthusiasm and some personal funding from the proceeds of his successful oil business. In Victor's words he figured out how to make a small fortune by investing a large one into a luxury-car biz.

He wore pinstriped suits and a wide grin, and he loved big toys. In 1987 Vic persuaded the Bond producers to put 007 back into an Aston Martin, following a sojourn that had seen him drive everything from a Citroen 2CV to a Lotus Esprit. Gauntlett's big personality led to him being asked to play the role of a KGB colonel in *The Living Daylights* . . . but he was too busy.

To clear up the confusion on the car used in the movie, *it was not a Vantage*. Vic loaned the producers his personal V8 Volante, a beautiful convertible with basically the same chassis as the Vantage but less power. Q 'winterized' the Volante by putting a roof on it, as well as wheel spikes, rockets, a tracking device and my favourite – deployable skis. This effectively converted the V8 Volante into the standard Aston Martin V8 known to aficionados as the Oscar India, and it would return in *No Time To Die*.

Prince Michael of Kent, on the other hand, did own a Vantage Volante. As the 1980s drew to a close he drove it to the Mille Miglia, where he spent the weekend with the ebullient Walter Hayes, senior public relations executive at Ford, and let him take it for a spin. Following that weekend, Walter had a meeting with Henry Ford II at his home in Henley. As they watched the boats bobbing along the Thames and sipped their coffees, the Deuce casually asked, 'What shall we do today?'

Walter replied, 'we could always buy Aston Martin . . .'

Dog Days Are Over

Walter Hayes was a grammar-school boy who had joined Fleet Street at the bottom and worked his way up through journalism and public relations, eventually becoming vice chairman of Ford Europe. He was a big-hearted man with large Eric Morecambe glasses that enhanced his wise-owl persona. He had a gift for timing a story as well as charming a superior with an idea.

When Hayes whispered into Ford's ear it wasn't a flippant remark, but rather based on a fair certainty that his boss and friend shared his appreciation of the unique qualities of Aston Martins. He also knew Henry 'believed that big companies could learn from small companies' and that Ford could use Astons for low-volume product development. An offer was made, and Ford bought Aston Martin in 1988.

Walter joined the board and approved the first all-new model in two decades. The styling was sleeker than on the previous V8 generation but retained the impression that it would still club a puppy to death while nobody was looking. Tadek Marek's V8 was deployed yet again, but modified by performance engineering specialists Callaway with four valves per cylinder instead of two to improve efficiency. It also had fuel injection to meet new emissions regulations. American laws had strangled the V8 down to a paltry 200 bhp, and by improving the exhaust emissions the new engine could run freely at 330 bhp. By christening the car Virage – French for 'corner' – Astons possibly hoped to convince buyers it might be able to tackle one.

There were acres of soft leather, opulent joinery, and, as always, the car was hand-built, but the Virage used simpler assembly processes to reduce costs. The chassis was based on another of William Towns' epic wedge designs from the Lagonda V8 Saloon, so it was well sorted. It was also given all the trappings that should have provided sharp, sportsmanlike handling such as double-wishbone front suspension, a lightened De Dion tube to brace the rear suspension, anti-roll bars and Koni dampers controlling the springs.

For actor Rowan Atkinson, a committed petrolhead who knows his onions, it was the new model he had been waiting ten years for. After parting with £120,000, he finally got his hands on it and was delighted by the air conditioning – provided by Jaguar – which delivered heat to the footwells in just ninety seconds from a cold start. The compliments ended there. In a 1990 *Car* magazine Rowan noted that the Virage felt 'limp' in power terms, with 100 bhp less than the previous V8 Vantage, while the suspension was 'wallowy'. He thought it the sort of car with which a 'New York lawyer could negotiate the potholes of Madison Avenue without loosening his fillings'.

The original suspension had been deemed too rigid for fat-cat private buyers, and was softened so much that when you turned the wheel, the body leaned lazily across without actually turning much. The result was an underpowered big girl's blouse, or as one customer put it: 'Looked like a race horse and goes like a cart horse.' Rowan hoped that the Vantage would right the wrongs of the Virage, but in the meantime he sold his and felt 'disenfranchised'.

While the Virage line carried on in the old-school tradition, a radical new concept was percolating that might save the company.

Walter saw that for Aston Martin to expand its appeal it needed a 'more accessible, superlatively sophisticated but user-friendly sports car which any reasonably competent driver could get into and feel immediately at home' – and cost £80,000.

Nick Fry was a young product manager at Ford's plant in Dagenham, where his team knocked out a Fiesta every fifty-two seconds. Nick was summoned to Ford's HQ at its concrete complex in Brentwood for a meeting. This was a rarity, and Nick suspected that he was in for a bollocking. Instead he was met by Walter, who handed him a photo of a clay model for something called a DB7.

Walter paced about the office, fidgeting with his pipe and talking non-stop for an hour. He covered a range of topics from his personal journey at Ford, discovering Jackie Stewart, kick-starting F1's most famous and successful engine in the Cosworth DFV, bringing Ford to Carroll Shelby, shaping the GT40 programme and, time being of the essence, a new chapter at Aston Martin.

'Anyway, do you want the job?' he asked.

Based more on respect for Walter than any knowledge of the actual mission, Nick accepted.

He left the big Ford plant behind him and drove to Newport Pagnell, from where he would oversee the various production units of Aston Martin. Having walked up the creaking staircase of Aston Martin's 'global HQ' at Sunnyside, a two-storey mock-Tudor villa, Nick surveyed his new office. Excluding the desk, it was empty. Not even a light bulb or a lamp to put it in, or a phone to call a service desk that didn't exist. *What have I done?* he wondered.

'Go to B&Q,' Walter told him. 'Get whatever you need.'

Now that he had a lamp, Nick descended on the factory where the Virage was being made. 'It was the biggest mess you can imagine' and totally removed from the clean automated systems he was used to at Ford. The engine alone cost £19,100 to build, and the car took fifty-six hours to paint. 'Engineers worked on the right side of the road, the men on the left, and neither side spoke to the other,' he recalled. One third of the workers were at home on full pay since the order list was empty. A few days later it rained, and when the roof gave way all the technical drawings were ruined. There were no copies.

Project XX

Things were a bit more civilized over at Jaguar, which had also been acquired by Ford.

Tom Walkinshaw was a hard-nosed Scottish businessman, a former racing driver who had switched to building an advanced automotive empire. He had drawn in the brightest minds and ruled with the rod to deliver outstanding performances in every series he competed in under the brand of Tom Walkinshaw Racing, winning the World Sportscar Championship with the Silk Cut Jaguar Group C car and the European Touring Car Championship with the Jaguar XJS.

TWR had been engaged to build Jaguar's XJ220 supercar, which had taken 1500 deposits on its launch. With four-wheel drive and a proven V12 racing engine it had buyers staring longingly at their Rolexes in anticipation of seeing one land on the driveway. Then Jaguar realized that a racing car would struggle to pass emissions tests in the real world, so they switched four-wheel drive to rear only and scaled the engine back to a V6 from an MG Metro rally car that rattled like a tin can inside your eardrum. Buyers' eyes watered further when the price increased to £470,000 for half the car they had originally ordered. Production numbers were reduced, and this freed up capacity at the TWR factory in Bloxham for building a new model.

Tom wanted to create a successor to the XJS, something cheaper and lighter that would accommodate the powerful V12 he had developed over years of racing. Jaguar invested in the project, which was assigned a different code name on a daily basis. Starting as the F-Type, it became the XJ41, Project XX, the NPX... Tom had hired a designer called Ian Callum in 1990 to oversee it. Ian had been working at Ford but was struck by the niche professionalism at TWR and threw all of his creative agility at the design.

The XJS chassis would form the basis of the new platform, but from the beginning Tom knew that Jaguar would not be able to afford to manufacture it. He told Ian, 'We'll pretend it's a Jaguar for now, but it's gonna be an Aston Martin.' When Jaguar rejected Tom's proposal, as expected, he linked up with Walter Hayes and carried on. Walter wanted to use Jaguar's AJ6 straight-six-cylinder engine in keeping with Astons of the past, an inherently tall unit that resulted in the distinctive bonnet bulge of Jaguars of the era.

Ian, the new boy, was given absolute autonomy to style the car. He had no intention of creating a 'Jag in drag'. There were to be no bulges; the engine would have to be lowered in the chassis. Beyond styling, there would be clear handling benefits from a lower centre of gravity. The seasoned campaigners at TWR told Ian where to stuff that overly complex idea. Walkinshaw dragged everyone in for a meeting, placed Ian in front of the old sweats and said, 'Ian, tell 'em what you want.' Down went the engine. Callum's styling was essential for the success of this project, and Walkinshaw knew it.

On another occasion, Tom came to Ian with a question. 'What's a PA?' he asked with a scrunched-up expression.

'It's a product approval, before you get to secure funding for further development.'

'Oh, right. What d'ye need for that?'

'You need a clay model to show the design, maybe some interior works and a business case for the board to go through and sign off or not.'

'Right. What's better than a clay model?'

'Well, you can do a fibreglass one and have it fully painted, which would look better.'

Tom nodded and walked off. Then he stopped and Ian knew what was coming next. 'What's better than a model?'

'Well . . . a driving one, I suppose.'

'That's what we'll do then, aye, really.'

In just three months Ian and the TWR team turned round a working prototype of the DB7, installed one of their V12 engines and filmed it driving across Tom's estate from a helicopter. Then it was driven into the showroom where the Ford board were meeting and their jaws hit the floor. It looked so good they thought it was a finished product. The next stage of funding was approved.

Wild Thing

Walter Hayes, now chairman of Aston Martin, was beginning to sweat a little. While stoking developments up at Bloxham he also had to keep sales alive at Newport Pagnell and make sure Ford didn't lose interest in the fledgling business. Aston sales represented just 0.007 per cent of Ford's total.

He wrote to the Deuce and explained that while his contemporaries had wondered 'Why I would risk my reputation trying to bail out a sinking ship, I don't value my reputation so highly that I would hesitate to do what I can for a great marque.' He was playing for time, and when his great ally and friend Henry Ford II died a few months later, it had almost run out.

The Virage was a motorized sofa, and buyers were doing anything they could to squirm out of their £20,000 deposits. Walter needed a high-performance Vantage version to get the punters in, and it needed to be an old-school behemoth in order to cover his bases in case the DB7 didn't cut the mustard. So while TWR cracked on with developing the DB7, the in-house design team at Newport Pagnell went to town on project Vantage.

They took the 5.3-litre V8 from the Virage and plugged in a pair of Eaton superchargers, then gave it a new crankshaft with Cosworth pistons and an engine management system by Ford to raise horsepower from 335 to a meaty 550 bhp. But according to Astons' engine chief Arthur Wilson, 'There was some concern over the social acceptability of such power at the time, so the quoted 550 bhp was conservative.' Let's call it 600 bhp, flinging you to 60 mph in 4 seconds and all the way up to 190 mph. The tyres were widened to 285/45 Goodyears to get more rubber on the road, and you could hear them coming from a mile away. The body was far more aggressive than the Virage, with broader wheel arches, bonnet louvres and shark gills trailing the front wheels.

The Vantage was an unruly thug at its finished best. In its pre-production form it was a malevolent swamp creature, and developing its handling was no mean feat. Walter got his old friend and triple F1 Champion Jackie Stewart onto the board of directors and hired him as a driving consultant.

When I first met Sir Jackie some years later, I was a young driver for his racing team. On race day the team manager made sure that the preparation bays were cleaner than an operating theatre. Jackie arrived with all the presence of royalty, immaculately dressed in pressed tartan trousers and tweed jacket. His eagle eyes scanned the cars, team, the polished racing transporter and then dropped to the truck's tyres. Goodyear was a sponsor, and its logos were painted white to make them stand out. The problem, Jackie pointed out, was that they were not perfectly aligned. The truck was lifted off the ground and all eight wheels rotated until GOODYEAR was in the correct 12 o'clock position.

Jackie left no stone, or wheel, unturned during his career: an Olympic-standard shot, a triple Formula 1 world champion and a driver with unrivalled finesse who took his F1 drivers to task as a team owner if their driving style was rough around the edges. Jackie could tiptoe a Tyrrell F1 car round the streets of Monaco at 170 mph using delicate inputs of the wheel and precise throttle control, and steering the Vantage was like arm-wrestling a hyena, so back to our narrative . . . In 1991 it would be interesting to see what his gimlet eye would make of Aston Martin's newest car.

To offer the driver an escape clause, the Vantage came with ABS braking for those eyes-closed moments of terror lying around every corner. Its colossal fourteen-inch AP Racing brakes were derived from a Group C Le Mans car and were the largest ever fitted to a street machine. The Vantage had savage power, and in a bid to stop the tyres spinning all the time, Astons' engineers had fitted stiffer springs and anti-roll bars. This led to an ugly issue called tramping, where the rear wheels jumped up and down on the road as they broke traction. This issue concerned Jackie enough for him to pen a letter to Ford's senior management team. A breakthrough soon followed. The front-mounted engine was transmitting massive torque along the drive shaft to the rear wheels and bending the differential. A huge torque tube was added to lock the two ends of the drive train together, and this solved the issue to a degree that was deemed 'acceptable' by Jackie. A sixth gear was also added to the agricultural ZF gearbox, which chattered away within an oversized tunnel running from the square gear lever to the boiler room in front.

Launched in 1993, this was the car my father brought home that white-knuckle day when I was just a lad.

They were a perfect match, and Dad cherished the dark green dinosaur grazing on the driveway above all living things. He would no sooner let me drive it than permit me to hold the TV remote. Passengers required approved footwear, and breaking wind aboard the leather land yacht was forbidden. But the Vantage held a dark secret . . . *It was hand-made.*

One winter's day the thud of extra-heavy feet at home suggested all was not well. The headlights on the 'f***ing Aston' had steamed up, shortly followed by the speedo. One morning the dashboard fell onto his lap. On the other hand, my girlfriend, now my lovely wife, accidentally burned a hole in the felt ceiling with a cigarette and not only survived to tell me, but apparently was told he 'couldn't care less'. That's when I knew that he would let me drive it.

Dad's record time from the farm to Tiverton Parkway train station was twelve minutes with a Honda NSX. His business partner didn't enjoy the journey and threw up on the platform. If I fancied my chances of beating his record in the Vantage I had my work cut out. The two tons of metal felt like three and filled the Devon lanes. On the open road the acceleration was unhinged. There was no turbo lag on the superchargers but for some reason the power came in unevenly. Going downhill into a damp section beneath some trees, I entered a long left-hander and applied some throttle on the way out. The front end slipped a little; I added some steering, then *whoosh* came the torque and I wished I'd worn brown corduroys. The rear stepped out and the rampant weight-shifting was like being on a see-saw with a really excited fat kid on the other end.

There was only one way to drive the Vantage, and that was with respect. The motor emitted a wondrous shriek at high revs, a useful alarm bell because the Vantage could spin its wheels in third gear on a dry surface. The animal force, combined with Rolls-Royce luxury, resulted in a total loss of sensation of speed. You saw the roundabout ahead and imagined your speed was about right, but when you checked the gauge, it still read 95 mph, and there was more braking to do. Despite the clever engine management system, fuel consumption barely reached double digits. Even with the giant brakes, you had to really lean on the pedal to stop it. The titanic torque gave you pulling power in any gear.

In short, I loved it. You could even drive it tenderly around town without drama and attract looks from the populace that were reserved for rock stars. The Vantage was the pyre a Viking would choose for the flaming journey to Valhalla – if it didn't break down on the way there. After one final straw, Dad tried to offload his Vantage to the dealer, who raised his palms and shrugged. 'Well, it is hand-built, sir.'

The Vantage was peerless, and no car maker would have the balls to make anything like it again. With a price tag of £189,000 it sold a respectable 511 units. The Aston Martin Heritage Trust were rightly proud of the mighty Vantage, and wanted to showcase its vast engine in their museum. An example was lowered into a prime position, and, in keeping with its badass reputation, the pedestal simply collapsed under the weight.

The Battle of Bloxham

Meanwhile, over at TWR, the DB7 was coming along nicely. Tom Walkinshaw secretly wanted to buy Aston Martin. His engineering team at Bloxham was producing the car of the future, while the old factory at Newport Pagnell was stuck in the past with its niche beast. Tom wanted nothing to do with the Vantage, called it a safety hazard, and he didn't want anyone poking their nose into his business either.

Nick Fry was stuck in the middle and getting it in the ear for the slow progress on the Vantage. Board meetings were pantomimes; the real planning conversations took place in the corridor with Walter. Then there were Nick's weekly meetings with Walkinshaw, who sat at the end of the table like a rugby prop forward flanked by his scrum. Nick adjusted his double-breasted suit and decided that attack was the best form of defence. Tom ran off a list of demands. Nick vetoed them one by one. Exasperated, Tom eventually slammed his folders shut and stormed out, saying, 'If this is how it's going to fucking be, I'm out.'

Walter cooled Tom's rage over dinner and promised to have a word.

'Well done, Nick. You really pissed him off.'

After that Nick and Tom got on like a house on fire, their regular meetings going on for hours. Tom had completely invested himself in the project and was as determined as Walter to see it succeed.

Jackie appeared in Nick's office one Friday afternoon. He wanted to expand his role beyond caretaking the red-haired stepchild called Vantage, and since the DB7 was a running prototype, he wanted to drive it.

'Let's call Tom,' he said.

Nick didn't want to. Jackie *really* did. Nick called Tom while Jackie hovered in the office.

'*He's not bloody well coming here*,' was the answer.

'What's he saying,' Jackie asked.

Walter had described this sort of posturing as a 'competition for the world's most famous Scotsman', and Jackie wasn't about to settle for second place. Nick told Tom they were heading over. They climbed aboard Nick's Escort Cosworth, a car he had developed, and set off towards Bloxham.

Driving with a world champion in the passenger seat was worse than any driving test. Nick's palms were sweating; he felt unsure of his steering position, the pedals, left from right. Some slow-moving cars beetled along in front. Jackie checked his watch. The Escort 'Cossy' was a rally-grade rocket ship. Nick dropped down a gear and gunned it past the traffic. It felt good being back in control.

Jackie's hawk-like gaze almost certainly spotted it first, but by the time Nick realized he was approaching a central reservation it was too late to return to the correct side of the road. It was shit or bust. He kept the boot in and charged ahead while scanning urgently for a gap in the barrier for a return to safety. He found one and rejoined the legal side, using his indicators for good measure.

'Sorry,' he offered meekly.

Jackie stepped out of the car at the industrial park and stalked towards TWR Engineering. A security man had to let them into the building since Tom had sent everyone home early. The shop floor was empty, and all the DB7 prototypes were up on lifts, safely out of reach and with their wheels removed. Walkinshaw had won the Battle of Bloxham, but *och aye* it would be a long war.

PART SEVEN
1988–2007

DB7

DBs, GTs
AND BALLISTIC SUPER-Vs

DB7

The DB moniker had been absent since the David Brown era, the DBS the last to bear his initials. In the mid-1990s Walter and Nick agreed that the new pretender would benefit from being anchored in the lineage and went to see Sir David Brown at his London apartment to ask for permission. Having thought he was 'rather out of favour' at Aston Martin, Sir David was delighted by the styling of the new model and readily agreed to DB7.

Even though progress was being made with both key models, there were some cold feet at Ford and rumours of selling Astons began to circulate. Walter drew up a fully fledged sales prospectus with a list of potential buyers that he plucked from the *Financial Times* and passed around the boardroom, bluffing unabashedly about the raft of lucky investors who would be desperate to take over from Ford.

Walter re-instilled enough desire to thaw their toes at Ford until the DB7 received final approval. To ensure that happened, he showcased it at the Geneva Motor Show, where it was universally praised and took so many orders that they had no choice but to make it.

Ian Callum's styling was what really sold the DB7, and though it was more rounded than any previous Aston, it possessed enough subtle notes such as the famous grille and the abrupt tail to single it out from the crowd. Underneath the hood, *sans* bulge, was Jaguar's 3.2-litre AJ6 lump from the XJS, a venerable power-giver in its standard straight-six form but boosted by a supercharger from 200 to 335 bhp. The problem was that building the cars involved a chaotic supply chain.

The DB7 underbody or framework came off the line of Jaguar's plant at Castle Bromwich as an XJS, then went to Coventry to have its front and rear overhangs chopped off and the main body added before going to Rolls-Royce for painting, then it went to Bloxham to meet the engines coming in from TWR for final assembly. Far from a mechanized operation, the old XJ220 factory at Bloxham was basically a barn full of racing mechanics who 'fettled with things'. In the pursuit of perfection it was difficult to raise their output from two per week to the required three per day, but with perseverance, recruitment and training, Nick Fry and his team got there.

To complicate matters, the bodies coming out of Jaguar were far from flawless, as Nick discovered at a car wash when one filled with water and drenched his suit. Walkinshaw conceived the idea of using composite materials for the bonnet and the trunk to save weight, although this complicated the process of getting a class-A finish on the paintwork. But the hard work was worth it.

Journalists lapped up the first finished DB7, Jeremy Clarkson saying the styling was so good you couldn't walk away from it without turning round for a second glance. Forecast sales of 625 per year were doubled to 1200, and Aston Martin was finally in the volume trade. The DB7 created a blueprint for a bloodline that continues to wow owners of the latest DB thoroughbreds. Once again, Aston Martin was saved from extinction by a few good men who would never say die.

Heart of the Ship

We had just installed the V12 in the first DB Vantage prototype and everyone knew we had a winning product on our hands.

Anthony Musci, product design engineer at Ford

Aston Martin entered a new golden era with two siblings that couldn't have had less in common. The wild-child Vantage that wanted to headbutt the governess marked the end of an era and was discontinued in the mid-90s. The purring DB7 that ordered its Martini shaken, not stirred, was a bit mild for my taste, but that was rectified by the DB7 Vantage edition with its bold advance from the 3.2 supercharged straight six to a very special 6-litre V12 engine.

Journalists blithely observed that the Aston V12 was 'just two Ford Mondeo engines taped together'. They could have used creme eggs for all I care, because the result was astounding, but the myth should be debunked. As Bentley proved in everything from aero engines to his long-serving straight six, nothing in engineering is wasted, and progress is a continuous process of refinement.

Porsche initially designed a lightweight yet highly powerful 2.4-litre unit with maximum durability that would inspire its first water-cooled flat six for the 1996 Boxster. Ford acquired the principle design and developed it into the 2.5 Duratec V6 engine for the Mondeo. When it was decided to create a V12 for Aston Martin, the engineers at Ford experimented with three sections of the Duratec welded together to test the architecture – this test mule was shown to journalists at the 1993 Geneva Motor Show with the weld lines exposed. The story was born of welded engines, but they should have counted: 2.5 plus 2.5 only equals 5.0. *This was not the real engine.*

A new magnesium alloy had just been invented – ZC63 – lighter and a third stronger than anything before, and a proto-type motor was cast from this special material. The proven piston assemblies and valve-train system from the Duratec were adopted as a baseline for the new unit with plenty of power-growth poten-tial. The official goal was to reach new levels of low-end torque as well as high revving power in a reliable form. Off the record, according to Anthony Musci who was undertaking the architec-tural research into the project, and hidden even from Ford's senior management, there was a secret goal baked into its design: 'From the start, the V12 would be designed to compete at the Le Mans 24 Hours. We knew if we did a good job, the desire to take it racing would come,' explained Musci.

Clearly whatever they produced would be measured against both Ferrari and GM's finest, and it would have to win. To assist the development team to fly under the radar they used new computer-aided design technology and simulation software to reduce costs just enough to remain within a research budget. But they would have to break cover when it came to production, which would require a specialist.

They approached Cosworth engineering, who offered to carry the entire project through final development and manufacture, intimating that the Ford team probably couldn't manage it by themselves. Their offer was rejected and the project continued step by funded step. Cosworth were only commissioned to carry out the final assembly and calibration process.

Higher-grade aluminium castings, a deeper block, thinner walls, stronger bearings and a proprietary combustion chamber with advanced inlet ports to boost bottom-end power resulted in a higher compression ratio that marked the 6-litre AML V12 out as a completely new and revolutionary unit. In September 1995 Ford's performance engineering team in Michigan finally felt they were ready for a first firing. Months of planning and adjustments had taken the engineers to the moment of truth where fuel would be added to the metal for the first time, followed by that initial spark. Issues with internal clearances, heat expansion, cooling and any number of electrical faults might reveal themselves as a puddle on the floor.

After long hours of final checks, the wiring loom was attached to a positive connection. Ford's technical boss, Bob Natkin, listened in on the phone from a press junket in Le Mans. They fired the starter. The engine barked to life, and the roar crackled down the line as the V12 called out to its earth mother from across the Atlantic.

Engine number 2 was sent to England and installed in the first DB7 Vantage prototype. A Visteon electronic management system processed over a million engine commands per second to control performance and efficiency. The unit delivered 420 bhp with plenty more in reserve, while 0 to 60 mph took just 5 seconds, and the Brembo brakes brought you back again just as quickly. A healthy improvement from the standard DB7, but there was much more to come.

What's in a Name?

The early 2000s were a twilight zone in terms of Aston Martin branding with the V-cars merging into the DB line and other shenanigans, but the good news is that from here on there would only be great cars to differing degrees. If I seem critical at times, it's because I have the benefit of hindsight from a pampered modern perspective. Lightweight construction is now everywhere, from the panels to the chassis they bolt on to, and engines are positioned low and inboard of the front axle for better cornering rotation.

Meanwhile Astons transitioned towards two distinct lines of tauter, more powerful cars that vary in their dimensions but follow the same engineering principles.

Ian Callum sums up the overarching philosophy: 'If a car is not absorbing the energy then you take it as the driver.' The hallmark of Aston Martin from now on would be delivering that raw energy without wearing out the driver.

Logic insisted that the next in the DB line should be called the DB8, but this designation was skipped due to fears it would mislead people into thinking it had eight cylinders, not twelve. At one point a young marketing bod sauntered into the canteen at Newport Pagnell, where the works department continued the restoration and building of heritage models. Some older buccaneers drinking tea from personalized mugs asked him what the new car would be called.

'Virage.'

Tea was involuntarily atomized as soon as the dread word was uttered. (This brain fart did, in fact, crystallize into reality some years later in 2011 as an in-betweener DB9 and a half . . .)

More importantly, Ian Callum had his magic pen in hand to draw the next generation of wholly un-Jaguar models. There were two distinct lines, with the V12 engine powering the bigger 2+2-seater grand tourer and V8 motors in the smaller two-seater sports car. So regardless of name-calling, if you want to know which line a model truly belongs to it's best to count the seats and the number of pistons!

Ian produced a shorter and more compact design called Project Vantage to house the V8, and in doing so introduced the DNA of subsequent Vantage models running through to today. One break with the past he proposed was for the engine to be mid-mounted in a direct challenge to the likes of Porsche and Ferrari.

Dr Ulrich Bez had taken over the management as CEO of

Astons and was determined to continue the firm's advance towards mass-market sales. He was firmly opposed to a mid-engined Aston and squashed that idea on his second day in office. He also decided to deploy the V12 instead and call the new model the Vanquish. It was designed fully in house using a Lotus extrusion technique of bonded aluminium with carbon-fibre crash structures to provide a rigid chassis. The body was similar to the DB7, but more lithe and defined.

The Vanquish first saw action in James Bond's hands during *Die Another Day* on the frozen Lake Jökulsárlón in Iceland, a low-friction paradise of icebergs surrounded by glaciers. The bad guy was driving a Jaguar XKR that Ian Callum had designed at the same time as the Aston. In street spec both cars were closely matched for weight and speed except that the Jag used a supercharged V8. In order to cope with the ice they underwent significant surgery under the hands of special effects coordinator Chris Corbould.

Chris had joined the real Q Branch behind 007 on *The Spy Who Loved Me* back in 1977 and helped make the modifications to the Volante required for tackling the slopes and ice lake of the Austrian alps in *The Living Daylights* ten years later. It taught him a valuable lesson, which was that 'ice lakes are unpredictable'. Engineer Andy Smith, of Batmobile or 'Tumbler' fame, led the modifications of both cars and in the interests of traction and speed added a front axle from a Ford Explorer to make them four-wheel drive. The tyres were also fitted with spikes to provide nearly as much grip as on a dry road.

*

The problem was that the ice was melting. Action unit director Vic Armstrong drilled the crew to get them off the ice in four minutes, which was an astonishing record for all the departments carrying their assorted clobber. Stunt driver George Cottle explained that the 'view through the windscreen was this bow wave, like a wall about fifty feet in front from where the ice was sinking under the car's weight'. Chris had the cars fitted with automatic inflation bags so that if they fell through it would keep them afloat.

The boys drifted in and out of the gaps between the icebergs for weeks of filming and on their last day the ice melted. The crew were safely evacuated and the lake was sealed off by Health and Safety.

The Vanquish was a decent motor but it still lost to the Ferrari 575 in a knife fight. The automatic gearbox was lazy on the upshift and made you look uncool as each gyration took you by surprise, and it didn't necessarily obey orders on downshifting. You needed the manual version to get the most out of it. The suspension was easy-going and soft, so it tended to roll across the rear axle, taking weight off the inside front wheel and creating understeer, which was a bit tame when more violence was needed. It was also £60,000 more expensive than the DB9, which would sell alongside it, for what was a similar car with exactly the same engine.

Lessons from the Vanquish led to the simplified VH aluminium chassis structure for the DB9. This was usable across other models, thereby reducing costs. V for vertical meant it could be used up and down the Aston range, and H for horizontal signified it could be used across Ford's cars as well. This chassis became the core of the future DB line, and every subsequent model would improve incrementally.

DBR9

The DB9 chassis and engine combination was begging to go racing. David Richards, known as D-R, was the chairman of Prodrive and had been campaigning the Ferrari 550 Maranello in GT racing, his team winning the category at Le Mans in 2003. With his mellow beard and shoulder-length silver hair, D-R was a smooth operator who looked like he could have fronted Status Quo if he hadn't been running a successful engineering firm. The 550 was remarkably similar to the DB9 with its front-mounted V12, and as that racing programme came to a close, David was looking out for a new challenge where he could apply his experience.

He discussed his ideas with Dr Bez and subsequently created a new company under licence called Aston Martin Racing. The purpose of AMR was to create a racing car out of the DB9 to enter in top-flight international competitions such as GT1. FIA rules dictated that the cars in this category were loosely based on their road versions in so much as the chassis, body dimensions and engine layouts bore some resemblance to the original. In truth the cars metamorphosed into a completely new species.

Metal body panels became carbon fibre, which were sculpted for vastly superior aerodynamic performance based on computational fluid dynamics calculations, a software substitute for a wind tunnel. Don't let the headlamps fool you; everything in between and around them had more in common with an F1 car, and certainly the rear wing, which stood out like a barn door. Giant Brembo six-pot callipers squeezed the brake pads onto carbon discs to pull the maximum force out of the fat, slick Michelin racing tyres.

Many of the Ferrari 550's most advanced features, such as its double wishbone suspension, springs, dampers, brakes and geometry were pulled across to the DBR9 and refined to work with the British car's bonded aluminium chassis. The Ferrari had used a traditional monocoque. The Aston's engine was lowered and moved inboard towards the driver to improve handling, and it also underwent significant modifications. The bore of the cylinder grew from 89 to 94 millimetres, and Prodrive designed new crankshaft, bearings and pistons to advance the output to 625 bhp with a 0 to 60 mph of 3.4 seconds.

Racing driver Darren Turner was with the programme from the beginning and has probably logged more laps behind the wheel of an Aston Martin than any human being. 'The GT1 [as the DBR9 became] was a beast to drive and my favourite racing car to date. It had it all, sequential gearshift with heel and toe, and that engine was very special.'

So was the opposition. Since the year 2000, GM's Corvette had been the dominant force at Le Mans. It was a sophisticated race car, and the team's formidable driver line-up extracted every ounce of performance. Developing the DBR9 to match it was a long road and took place far away from the glamour of Le Mans.

Having won the 12 Hours of Sebring in 2005 under the glorious Florida sunshine, Darren and the other drivers spent four days doing back-to-back twelve-hour endurance testing at Elvington, a drab airfield circuit in Yorkshire. With no real corners on the site, the team marked out a soulless coned course, brimmed the fuel tank and sent the drivers out one by one to drive around until their arms fell off.

'Tomas Enge was fastest, of course,' recalled Darren with a resigned tone that would surprise anyone who didn't know Enge, a Czech driver with a sympathy for machinery comparable to Rambo's relationship with bullets. No kerb stone or piece of tarmac was immune from his rigorous pounding, and certainly no rubber cone was getting between him and the top of the timesheets.

By day four the mechanics were as bored as boulders, the sign of good reliability. Knowing that cone-killer Enge was flying out of Manchester that afternoon, they organized a farewell gift and wired his road car's brake pedal to the horn. Karma spoke to him for the duration of his trip to the airport through a series of head-turning toots at every braking point.

24-Hour Sprint

In the early 2000s Le Mans was a very different place to the dirt country roads that Sammy Davis had sped along with stones peppering his mudguards. The course had been tarmacked before the Moss era, and today the long Mulsanne Straight is split by two fast chicanes. The Porsche Curves have replaced the plunge past White House, and the run to the finish line was slowed by the

Ford Chicane. Yet the rhythm and the soul of the place remains.

In the old days at Le Mans you might have eased off on the brakes if they felt like they were getting hot, or back off to spare engine rpm. Teams were afraid to run at full pace in case it hampered reliability. Not now. These days you don't spare the car at all. During a qualifying session a driver aims to create a single masterpiece that presents the ultimate fastest time the car could produce. At Le Mans that is expected every single lap of twenty-four one-hour stints. At the end of a stint you pulled into the pit lane at full speed, braked hard down to 60 kph and toggled a switch to hold your speed at 58, eyeballed your crew and swept into your marked square until the nose kissed the lollipop board, and killed the motor.

Your body relaxed for thirty seconds as the fuel hose rammed home, the car rocked its way up on pneumatic stilts and the air guns whizzed at the wheel nuts. Time to breathe deeply, oxygenate, let the blood flow into the tense muscles, prime the pump to the fresh drinks bottle piped to your mouth to make sure it installed properly. Get ready, clutch in, ignition on, *thud* as the car hit the ground, fire it and go. You felt a little cooler, a little more energized, and if you were lucky you had some fresh tyres to play with, but that was usually only once every three hours.

In attack mode the first right-hander was taken easily flat out, without slowing, but using every inch of road as far as the gravel on the far left to take a clean line into the right and land square for the hard brake for the Dunlop Chicane. *Flik-flak* across the choppy blue kerbs and a blind exit before diving downhill to the Esses, another hard stamp on the brakes, carrying speed into the left with a heavy steer to drag the nose in, balancing on power into the right and squirming away over the crest.

A short run to Tertre Rouge, brake very late, every inch counted but don't touch the outside kerb on the left or it pulled you off the road, turn the wheel bravely so you got enough rotation to hammer down early onto Mulsanne. Two chicanes here, both similar but one to the right and the other left. Brake eye-wateringly late, so late that if you didn't release the pedal you would never make the corner, the car whistled in on edge, you changed direction and crept out minding that bronco tail.

Braking for the tight Mulsanne was on a kink that lightened the front right wheel. The slowing force might lock it up, burning a hole through the tyre flesh and possibly causing a puncture. A puncture necessitated a stop and a note in the log from someone like Wyer that read, 'Unscheduled stop for self-inflicted tyre damage. Collect P45?'

You felt for the wheel to twitch, adjusted the brake, turned right and climbed the apex kerb, tickled the throttle, and rode out the vision-blurring bumps at the exit. The long road to Indianapolis was good respite. Poke a hand through the narrow window to cop some airflow, breathe, enjoy disobeying the farcical white road markings shooting underneath the chassis as you ran up to 186 mph.

Study the mirrors for the predatory prototype-class UFOs that came out of nowhere and shot past with a plus speed of 50 mph. I used to be the poacher racing those 800 bhp monsters before becoming their prey in the GT class.

You lined up on the left for the world's fastest corner at the tree-lined right into Indianapolis. LED lights showing rpm shining a full bank of green, yellow and red towards your visor as the motor tapped out, 186 mph. You carried maximum downforce, pressing the tyres into the ground, and the car was not capable of sliding gracefully on the limit any more. It would either stick to the rails or twist towards the trees and someone had to write to your folks, 'He was an optimist, we'll miss him.'

Hold her steady and lift off deep into the bumps, barely slowing into the right without eating all the exit room because you had to stop, straighten it up, down to second gear, and switch left into the banked apex and then away to Arnage. It was bumpy on arrival there and easy to lock the rear tyres with a shoddy downchange – you heard the motor die if you really screwed up – so you kept it neat, and blasted off towards the fun zone.

The Porsche Curves were a continuous roller-coaster stream of fast, connected curves. The first right was entered after a light brake, holding it tight to the inside to set you up for the next left. In front of you was the welcoming wall, and the trick was to take this one without lifting off the gas. In the GT class this was dicey but doable, and every time you managed it, gave a fresh burst of adrenaline. The extra speed it gave you added just enough additional down-force for a lift only into the downhill left with the welcoming wall closer again to the outside. The rails you rode on quivered until you eventually braked with the rear wobbling for the tortuously long right-hander, into the final left and on to the home stretch.

The race organizers had installed some tasty red kerbs for the fast chicane to reduce speeds around the pit entry. It hadn't worked, and you could pile into it in third gear, land in a heap on the other side and collect yourself for the slow turn on to the pit straight ready to try it again.

And breathe.

That was my experience on the practice day at Le Mans a few years ago as I acclimatized to the Ferrari GTE, having missed the pre-test earlier in the year. I climbed out of the lurid lime-green machine, the colours of Krohn Racing, and went to review time charts with the engineer.

'How hard are you pushing?' he asked ominously.

'I'm giving pretty much all of it,' I answered.

'We're way off the pace. Two seconds.'

My guts twisted into a knot. I felt as useful as a fish riding a bicycle.

The circuit TV showed live footage of the track, and when I watched Darren Turner go through the Ford Chicane in the Aston my jaw hit the floor. White lines bordered both sides of the tarmac and delineated what was 'ours' from the greenery that was 'theirs', meaning the officials.

The rule was that you had to keep at least three wheels on the black stuff. D. T. was so far across the kerbs that his outside wheels were barely touching tarmac: in fact he was airborne. In a masterful display of acrobatics, the Aston's wheels gyrated in their arches, and he landed it beautifully for the next turn at a cocked angle. A few minutes later one of the Ferraris crashed heavily there. I followed D. T.'s progress round the lap, and he was over the white lines everywhere.

It turned out that the officials had abolished track limits and installed a jungle of kerbs to control us using the laws of Darwin. I found the lost lap time using the extra room, and we traded sweat with the Aston Martins during the race, with a shot at second place until the gearbox made other plans.

That was Le Mans for you. You drove at 100 per cent and hoped the car held together. The driver's job was to maintain the fitness level of a marathon runner to sustain extraordinary levels of concentration without succumbing to physical or mental exhaustion, and keep threading the needle. Without a doubt the most reassuring thing a driver can have is the support of a team focused on winning, pulling as one. You can level up.

The staccato sound of an exhaust banging against the rev limiter like a machine gun lifted my gaze down the pit lane as Turner cruised through in the Gulf-liveried Aston. He steered into the welcoming arms of the AMR crew, was barely stationary before the fuel splashed home, the door was open, the next driver yanking him out, door slamming, and gone. It was a perfect pit stop, monitored from the pit wall without even a nod of satisfaction, just some notes through the head cans. This was a team en route to victory in 2014.

Well Done

For Aston Martin Racing it was an old bug that reared its head – the race car that wanted to cook its driver. Back in 2005, Le Mans took place during an especially hot summer. Imagine wearing a thick balaclava, helmet, leggings, long-sleeve shirt, triple-layer

Nomex suit, boots, gloves. All compressed into a moulded seat that eliminated the air circulation needed to evaporate your sweat and cool your trunk, with just an air duct blowing warm kisses. Inside the DBR9, the cockpit temperature rose to *seventy-six degrees Celsius* before the gauge broke. The rear screen literally melted.

During the race the drivers felt increasingly spaced out, and Darren Turner wasn't himself. He clipped a bollard going through the Ford Chicane and had to come into the pits to sit still and serve his penalty time. As the cooling airflow stopped, the heat soaked in. He felt like he was being strangled. When Darren left the pits he was so dizzy that he missed the chicane on Mulsanne and brushed through the gravel trap. Somehow, he completed his stint.

While Darren was plugged into an intravenous drip in an attempt to rehydrate him, David Brabham took over for his session in the microwave before handing over to Stéphane Sarrazin. By the close of the race and despite receiving multiple IV drips, nobody was fit to drive. Sarrazin had double vision; Brabs couldn't walk because his calf muscles were in spasm and torn to bits, while Darren's feet were raw with blisters. Fortunately a radiator leak caused enough of a delay in the pits for Sarrazin to see one of everything, and he finished the race.

This wasn't a question of driver fitness. When the car came in at the end of the race one of the engineers climbed inside with his computer to download the data and the penny dropped.

'George,' he shouted to the team principal, 'come and feel the temperature in here. It's bloody unbelievable.' It always was until you felt it for yourself. George Howard-Chappell was a tall, strict disciplinarian whose primary focus was the car and running his team as a lean, mean fighting machine. He realized they had missed something. From then on the engine received significant heat shielding, and the drivers were equipped with cool suits. By 2007 this and all the other the bugs were ironed out, and the mean machine was ready to battle the Corvettes at Le Mans.

'We were expected to win,' recalled Darren. 'There was huge pressure, and you could see that on the boys' faces. The preparation was enormous, and they were beasting themselves during pit practice.' There was little to differentiate the combatants for pace, but fast pit work reduced the team's static time down to a remarkable eighteen minutes out of the twenty-four hours. The Astons had an edge in the dry conditions, and by the end of the race were a full lap ahead. However, a big black cloud was hovering over Le Mans, and David Richards, working as a weather man in a helicopter, reported that it was 'absolutely pissing down' at his location.

Brabham had done his share of stints and was rather enjoying the feeling of letting the other blokes end the race while he relaxed in dry clothing. Rickard Rydell was due to finish, but when the heavens opened and the windscreen wipers went into overdrive, he knew that someone with more experience of the DBR9 should take over. Brabs was reloaded.

For some reason the Vettes were much faster in the wet and just carried on. Brabham tried to follow but was aquaplaning badly. The water lifted his wheels off the tarmac to the extent that they were spinning in third gear on the straights, with every chance of him rotating into the wall at any moment. When the lead big yellow Corvette drove past him to unlap itself he thought it was only a matter of time before it caught him again and took first position.

'The race had been faultless until then and suddenly there was this enormous pressure of maybe blowing it. My body was shitting itself and I just had to think, *trust my hands*, my hands will get me through this,' recalled Brabs. Racing in the rain was a psychological battle, and you were safer driving dangerously with your mind ahead of the road than driving half-safe and reacting to the sound of every splash.

Brabham centred himself, found a rhythm and danced with the bucking bronco. As he took the chequered flag, the roar of the V12 was heard far and wide. *Born in Michigan, raised in Banbury, came home in first place at Le Mans* to score Aston Martin's biggest victory since 1959.

However, this was the swansong for Ford, as they had already decided to sell their stake in the business to a consortium of investors led by David Richards, who subsequently became chairman of Aston Martin. With the Le Mans win and a rock-star petrolhead in charge, the scene was set for a long-term commitment to top-line racing, which inspired broad international customer demand for racing versions of the DB9 across all levels of motorsport, ranging from the ultra-fast GT1 through to the GT2, GT3 and the practically road-standard GT4, which was basically a DB9 with a roll cage. The tide was rising for the road cars too.

The Stig

Top Gear, that motoring programme on the telly, was a bit like an extended stag do albeit built around the odd film about cars. The Stig, an anonymous racing driver dressed in white, was always looking for an opportunity to unleash the full potential of everything he drove. His mission was pure, unbiased performance, and as the FIA GT3 series got into full swing the worlds of fiction and reality collided.

The GT series featured twelve different supercar manufacturers, all with different engine and chassis combinations. To equalize the competition, the FIA organized 'balance of performance' sessions at the Paul Ricard circuit, situated in the lush countryside outside Marseilles. The track had been modernized, and instead of gravel traps it was bordered by expansive tarmac areas painted with increasingly abrasive blue and red stripes, like a doodle by Picasso. FIA-appointed drivers would test each car, and the data was pooled to see how fast they *really* were. If a car was too fast it would be restricted either on its engine, by carrying ballast or raising its ride height, the last thing you wanted. So, for the purpose of these tests, the teams were keen to look good, but not *too good*.

Top Gear magazine had scored a real coup and managed to arrange for me, as the Stig, and an Aussie journalist called Bill Thomas to drive every car on the grid. In racing circles this was unheard of. I knew this world well, having driven for Ascari Cars' GT programme and scored a few pole positions for them, though the journey to the chequered flag proved more elusive. Dressed as the Stig or not, it was impossible for the Ascari lads not to recognize the familiar hooligan in Nomex, but they could be trusted to keep a secret.

The Ascari KZ1R GT3 was extremely capable due to its immensely stiff carbon-fibre chassis, which enabled it to corner on the proverbial rails. Stiffer cars are more lively on the limit, and there were sharp moments of correction as you squeezed the gas, and the mid-mounted BMW motor cleared its throat and shot you out. Not surprisingly, Ascari was hammered by the FIA, who put more lead ballast in the KZ1R than anything else on the grid and raised the ride height to slow it down. I managed to pop a time in at the top of the charts, and the other teams seemed to take notice.

Bill and I walked down the pit lane to the Audi garage, where the roller shutters were being closed in true Walkinshaw style. Bill's brow wrinkled. He was having the ultimate day trip away from the office, and in his innocence was hurt and surprised when the Germans fobbed us off with a bogus technical issue.

Ferrari was next, and the blood-red F430 Scuderia looked fast just sitting there, with its winged nose section, long low-slung body and a whopping diffuser at the rear to suck the chassis into the ground. The Ferrari mechanics were as enthusiastic as ever, attentive with the belts, informing you of tyre pressures and settings, espresso coffees on tap and making you feel like a million dollars at the centre of their universe.

You got a sense of the Ferrari just rolling down the pit lane; the raucous response from the throttle was instant, and the ride smooth but assured. Touching the brakes communicated a clear weight transfer to the front tyres and a turn of the wheel created a direct and measured response in the suspension. Mid-engined sports cars were always better on track, and the Ferrari was the king. It produced more grip and allowed me to brake deeper than in the Ascari, control the line and accelerate earlier. I was anticipating a ballistic lap time. Evidently, so was the team. Before I reached the last corner the car ran out of fuel, and I just managed to coast it back into the pits without recording an official time.

Corvette were reigning GT3 champions, and it was quickly apparent how they did it. GT3 regulations mandated that you had to run a pair of drivers of mixed ability – one pro and one amateur. The Corvette ZO6R was front-engined, like the Aston, and was so easy to drive that anyone could have jumped aboard and driven fast. It had enabled their nonprofessionals to achieve more pace than the amateur drivers of the trickier machines. They gave me enough fuel for two laps, and the car was planted and balanced everywhere, so it posted the fastest time.

We sauntered over to Aston Martin, where a sea of green filled multiple garages, and the professionals in matching team gear smacked of Formula 1. The cornucopia of delights there included the Big Bertha GT1 as well as their new GT2, 3 and 4 cars. Engineers with laptops went in one direction, public relations staff went the other to sort interviews and deliver urgent messages to the management team sitting on the 'prat perch', the covered control area on the pit wall. It looked expensive: big leather chairs, probably a jacuzzi just out of sight.

The GT4, or Vantage N24, was essentially a road-going Vantage with slick racing tyres and a roll cage. Besides being a safety feature, the cage added rigidity that enhanced cornering ability. It was a doddle to drive, although it felt a little unkind thrashing something that lacked the heavier-duty suspension elements of the thoroughbreds and seemed rather pressed into service. It was an easy-going entry-level tool for frightening oneself round the Nürburgring.

I took the GT2 next, which was in development to meet new regulations and subject to close official scrutiny. My suspicions of potential sand-bagging were confirmed by the reluctance of the engine to pull the skin off a rice pudding. I brought it back in and tried not to insinuate foul play while the engineer hid behind his laptop.

'It's really down on power, almost misfiring,' I explained without saying that he had obviously detuned the engine control map with that evil computer of his and turned a prince into a toad.

'Oh, really?' he replied with Oscar-winning incredulity.

I thanked him for the experience and moved to the DBR9.

From a distance it resembled a normal GT, but when you looked closer there were aero gadgets everywhere. The front splitter shaved the ground, scooping air into the classic Aston-shaped grille so that it flowed across a wing-like structure hidden under the bonnet. The flat floor let air flow underneath the car at high speed, creating an area of lower pressure than that interacting with the aerodynamic features above and creating the vacuum effect called downforce. The air met two features at the rear, the whale-tail wing on top and the venturi diffuser at the bottom, which sucked the car into the deck. It was more like an F1 car in sheep's clothing.

I climbed into the upright chair and surveyed the spartan

appointments. The sequential shifter rose vertically from the silver heat shielding on the gear tunnel to my right. You just yanked it back to change up the gears and banged it forward to change down. The lever was wired with an automatic ignition cut that sensed your pull and stole the spark to the engine for a fraction of a second, releasing enough tension in the drive train to shift gear without lifting off the throttle.

The steering wheel was bigger than the average dinner plate with a flat bottom and covered with velvety Alcantara faux-leather. No wood here. The head-up display was dead simple, with a large LED number to indicate what gear you were in; not as daft as you might think, given that you could lose count in the frenzied braking zones. The central console was the usual confusion of fuseboard and fire panel. Press that red button by mistake and spend a year in purgatory for hosing the engine bay and your nether regions with extinguishing foam.

The fuses were there to keep the electrical circuits alive by shortcutting faulty systems, activating secondary fuel pumps and reserves. There were various engine maps usually *verboten* to touch unless instructed by someone from the prat perch. All of this to memorize with your eyes closed so that at 2 a.m. while cruising down the Mulsanne Straight at 180 mph, your hand could find the correct button in the dark when you received a message on the radio to make a change.

Details like the intricate carbon fibre of the door, woven with the Aston Martin logo, raised the hairs on my neck. I was sitting in the mother ship. This was serious business for Prodrive, who had to finance the entire operation independently through a combination of commercial success and creativity. There was no money tree, and the pressure was always on.

If there's one thing a team manager hates above all things when he's following a tight testing schedule, it's the media. George Howard-Chappell had the presence of a headmaster, and his office was the car. He didn't know me from a hole in the ground, but he had regularly watched Clarkson using a sledgehammer and setting fire to stuff, so finding a *Top Gear* creature driving his £500,000 pride and joy probably disturbed his chi.

The number-one mechanic on the car received an instruction in his headset and told me to fire up the engine. It sprang to life and the revs shot up quickly in response to the throttle. Racing clutches can be snatchy, and to avoid looking like a numpty when you pull away, you lay on some decent rpm to avoid stalling. As I did this, the headmaster loomed into view with a long wagging finger. *Shit* . . . I had been caught eating jelly babies during choir again.

'*No revs, you'll fry the clutch.*'

'OK.' Embarrassing.

I departed the pits like the car was made of glass, then got down to business. Slick tyres and carbon brakes don't like being kept waiting as they only work properly at higher operating temperatures. The brake pedal felt wooden at first because it was cold and it didn't stop that quickly into turn one. The sound of the engine revving to 7,000 rpm was fantastic: even with ear plugs you could hear the wail of a snared trumpet and the staccato bang with each ignition cut from the gear change. The steering was light and precise and would harden as the tyres came in.

At the end of one lap everything was working, I passed the pits in fifth gear at around 165 mph and eked my way up to 170-ish before lining up on the right to tackle the cambered chicane. Bright red and blue kerbs marked out the track, and I was astounded how deep I could brake, almost in the corner. The carbon brake discs ate the speed; I banged the fat gear lever forward, down to second and turned uphill into the left. The ride felt as smooth and solid as a billiard table until I crossed the apex kerbs, which put an awful shimmy through the suspension. I wouldn't do that again. Picked up the throttle and crested the rise, turning right, and squeezed away again through the gears.

Anything in third gear or above deployed up to a maximum of 500 kilos of tarmac-squashing downforce, less in Le Mans trim, where you couldn't afford the drag along the big straights. It was wonderful feeling the G-force, carrying entry speed while using throttle inputs to subtly adjust the handling attitude between understeer and rotation. I could see why Darren loved it, and two laps went all too quickly. If I was being really greedy, I might wish it had a bit more power.

I did try the GT3 at Ricard but enjoyed a more visceral encounter during one of *Top Gear*'s many homages to World War II. It had come to our attention that another group of presenters had had the audacity to produce a car show, as if any besides *Top Gear* was necessary, *and* they were German. A competition was inevitable, so the gauntlet was thrown down and, as it usually does, it landed in Belgium. I reported for duty at the Zolder circuit at 0800 to help put together a little assault course for the three amigos, who were arriving later.

I walked up to our forlorn director Phil, who was recording some B-roll footage and trying to remain a quit smoker.

'Do you want me to try them out?' I offered.

He glanced doubtfully towards the miniature Panzer and British tank in the car park. They looked home-made, puny and ready to fall apart.

'Er, no. I don't think that's a good idea.'

'No worries. Can I go and try the racing car then?'

Besides overseeing a driving course made of giant Lego blocks and straw bales, my real purpose was to compete in a race against the opposing team's driver. It remained to be seen whether it would be a genuine cockfight or a bit of pretend. I was hoping to meet the team and see if the car was any good.

'No, you'd better stay here just in case.'

Film people kept you until the end of time 'just in case' they might need you, and then kick-bollocked and scrambled for the thing you were actually waiting for. At least this had the hallmarks of an entertaining shoot.

There were a series of points-scoring challenges through the day, and our go-anywhere weld-anything mechanic Steve had soldered together several pairs of cars to convert them into double-deckers. The man on top did the steering while the one below operated the pedals. Assuming such a race was survivable, there was then a gymkhana while being shot at by the little tanks, and finally Clarkson would race, or drive in the wake of, his idol and heroine of the Nürburgring, Sabine Schmitz.

The air behind us filled with a throbbing bass sound, and we turned to look up. They came into view above the trees, the stunning and familiar shape of three Supermarine Spitfires flying in formation. The proud nose, curved wings and the athletic frame covered with friendly camouflage and squadron markings. The hum of the Merlin V12 rose in pitch as they flew towards us then flattened as they passed overhead.

'Oh my God,' said Phil, then gurgled with laughter. I had seen Clarkson wrestle a megaphone away from a Tamil Tiger in Parliament Square to complain about high-octane fuel prices, but this was an impressive leap of political incorrectness, even by his standards.

The rest of the day went much according to plan, from a German perspective, and it looked like their beach towels would be the first onto the sunloungers back at the hotel. The grand finale was a two-lap race round the former F1 circuit itself, and the starting positions were staggered in accordance with the teams' points score, putting *Top Gear* a long way behind.

Germany's entry was Tim Schrick, a touring-car racer at the wheel of a GT3 Porsche 911 versus James May in our GT3 Aston Martin DBRS9. There wasn't a cat in hell's chance of James winning and, with tongue firmly in cheek, I appeared in his place wearing the Stig outfit, Sabine remarking that James had 'lost thirty kilos in five minutes'. Everyone was in on the joke, but at that point the script was put away.

'You *can* take him, can't you?' Jezza had asked me in the changing room.

'Yeah,' I replied with some confidence.

'Good.' He lumbered off for another smoke.

With ten minutes left in the day I finally met the French team running the Aston, by which time the director was having kittens because the circuit was about to close. Schrick was already on the track pounding laps, evidently not pretending, getting his tyres and brakes hot and deploying other useful tactics like learning where the track went. I jumped aboard the Aston, slid the seat into position and fired the engine. It cranked reluctantly and then a few seconds later hit the rev limiter without me even touching the throttle.

I switched off the noise and called the crew chief over. Two of our producers were on his shoulder in seconds asking me, 'What's the problem?'

'The throttle is jamming open,' I replied.

'We really need to get this car out there before they close,' one continued helpfully.

I explained the problem in pidgin French to the chief mech, who took a look under the bonnet, then shrugged in a Gallic fashion. To be fair he had been kept waiting all day; now it was 5.02 p.m. and 'A glass of red wine if you please' by his reckoning. The producers were pressing for a decision. While it would be bad if the car developed a mind of its own and accelerated me into a tree at warp speed, it never looked good to be the guy who flew all the way to Zolder to chicken out at the last minute. I tightened my nut straps and took her out.

Zolder had character, with some fast corners, a good flow and camber changes absent on more modern places. Halfway round my out lap, the throttle jammed open going into a hairpin, and I sped up when I wanted to slow down. For a moment I thought I was headed for the gravel trap and an early bath. I put the clutch in and tapped the throttle pedal, which seemed to clear it, then drove to the starting grid.

Gavin, our cheery Scottish assistant director, ran up the door. 'Do you need anything, mate?' he asked.

Schrick was a long way ahead in his nicely warmed-up Porsche.

'A rocket launcher?' I suggested.

'You can do it. See you on the other side.'

The grid was cleared and we set off line astern for the rolling start. Schrick passed the line and dropped the hammer, followed a few beats later by me. Turn one was a fast third-gear left, and he took it well. This wasn't going to be easy. We flew round a double right and then onto the back straight, which was my major concern as regards the sticking throttle because we were both doing 165 mph. The car behaved . . . for now.

By the end of lap one I had barely closed the gap and needed to be braver on the brakes or seek other employment. We shot past the pits for the final lap with all the presenters leaning over the wall. The adrenaline was really pumping, the taste of metal and an intuitive bond with the Aston urging me on. I braked as late as I dared into turn one and just as I turned in, the throttle went wide open, instantly pushing me wide.

I hit the brake, locked the front tyres and skidded over the exit kerbs until I sorted it and got going again. It lost two car lengths to my man in front. I muttered a prayer and braked late for the chicane on the back straight, clattered the kerbs and pulled Schrick within three car lengths. There wasn't enough space to do him at the next chicane so I waited for the downhill run to the final hairpin to deploy my foolproof plan: *Don't brake until he does*.

He left the door open on the inside, assessing, reasonably, that I was way too far back, and when his brake lights came on I lunged down the inside. The Aston was mega on the brakes, but I was still surprised it stopped in time for the corner, as was Schrick. He didn't have a response, and it was 1–0 to *Top Gear*.

The presenters were all smiles in the pit lane, and the next scramble was to get to our flight. Jezza spotted the memorial to our mutual hero Gilles Villeneuve, the F1 legend who was killed at Zolder in 1982.

'He was the last of the late brakers,' remarked Clarkson.

'The bravest of the brave,' I replied. I think Gilles would have enjoyed the DBRS9.

PART EIGHT
2008–Present

Valkyrie

NO TIME TO DIE:
RIDE OF THE VALKYRIES

Thanks to Ford's injection of technology and the stellar designs of Ian Callum, Aston Martin was both suave and ahead of the performance curve. Those early press reviews loaded with superlatives about the handling of Lionel Martin's lightweight cars could just as easily have applied to the twenty-first-century range. The big difference now was abundant, controllable horsepower. Being so close to perfection, we can expect Aston Martin to throw us a curve ball at any moment, and they will. But first, an old colleague clears his wardrobe of double-breasted suits, dons one tailored to his muscular frame and eases himself behind the wheel of Astons' latest thoroughbred.

Unbreakable

James Bond had returned to the big screen in spectacular style with Daniel Craig's visceral take on 007's duties for *Casino Royale*. The company car took a right hammering, being barrel-rolled a record 7¾ times when Bond swerved and flipped the car to avoid running over a damsel in distress. For the 2008 sequel *Quantum of Solace*, 007 was issued with a new Aston DBS and no penalty points were added to his licence to kill it.

The main chase sequence of the movie was filmed around Lake Garda where the stunt team was given an older DB9 for testing as well as *seven* Aston Martin DBS cars for use in filming, all of them brand new, smelling of cow hide and stunning to behold. My first mission was to take one of these £160,000, 190 mph supercars to my favourite airfield for a thorough stress-test.

Gary Powell was the 007 stunt co-ordinator overseeing all the fight training and driving sequences. A second-generation stuntman, Gary was born with a bag of stunt pads in his hand and the contents cringed every time it was opened. The DBS would be going through the mill in a fast-paced chase around Lake Garda, tearing along the mountain passes, squeezing through hand-cut tunnels, through opposing traffic and side-swiping anything that got in its way. And it would be doing it every day for three months. Gary's logic in asking me to push the car beyond the limit was that we needed to know its weaknesses.

I pitched up at Dunsfold aerodrome in Surrey, waved at the greying security guard, hopped over the speed humps, turned right past a knackered Hawker Hunter jet fighter and arrived at the apron of broken concrete beside the giant green shed known grandly as the *Top Gear* studio.

A brace of DBSs were parked up alongside a large stack of tyres and some familiar faces were setting out their stalls. Neil Layton was another convert to the Q-tribe, having run apes like me in GT racing and rally for years. I first met him at Silverstone for a round of FIA GT, just moments before the qualifying session when the Ascari GT car I was driving developed health issues.

Back then, from the driver's chair, strapped in with a helmet muffling anything you might contribute to the critical engineering problems of the day, you quietly watched the diorama and waited to be told whether the patient had died or was to be taken onto the track and have its neck wrung. Young faces full of panic ran around the awning looking for a magic bullet, fuel pumps, laptops, a Bible perhaps. Voices rose an octave until the top dog arrived and quietly took over.

'Right, stop. Bertie, put that down. Richie, get the engine cover off and grab the spare ECU.'

The wheels went back on, and I was lowered to the ground. Neil paced round to my window and in his usual lackadaisical tone said, 'Should be good now, Ben. See what she'll do.' The Ascari was fast enough to qualify on the front row. Sadly, while parked there before the race started, I heard a funny clunk and felt the car sag. A suspension arm had snapped, and even Neil couldn't knit a new one on the spot, so we were done for the day.

Back at Dunsfold on Bond duty, here was Neil as relaxed as ever: sunburned, wearing knee-length shorts, hands on hips, shoulders rounded, whimsically raising an eyebrow as he conducted his orchestra. On seeing me: 'Well, so this is what you call show business is it? Can't even get a cup of tea in this dive, I had to send a man out.'

He was happy with his motors though. The DBSs were painted in dark metallic grey. Sleek grand tourers with broad dimensions that were understated enough to separate them from an Italian sports car.

Some critics thought the DBS was a marketing gimmick, just a rebadged DB9 with the same 6-litre V12 engine and chassis as the old model. But as with the DBR1, the secret to perfection was continuous refinement of the core elements. The V12 was the pinnacle of power plants, but by tweaking the inlet ports to let the combustion chamber breath more air, power rose from 450 to 500 bhp. The 'superficial' body tweaks included a carbon-fibre splitter at the nose to create downforce and stronger shoulder profiles that channelled air to generate aero grip. The widened track and lowered ride height improved stability, pulling the car reassuringly into the road with subtle tweaks to the shock absorbers that kept you there.

The DBS felt more agile and responsive at the wheel. Any vagueness from the old Jaguar DNA was gone. Another thing that stood out was the braking performance. The software controlling the ABS had been modified into a more aggressive version of R2-D2, and when you applied maximum pressure to the brake pedal it worked intelligently with the tyres to gnaw at the tarmac and shed speed.

I climbed aboard and breathed in the scent of new car

emanating from the acres of stitched leather. The helm was perfect in the DBS. Your hands aligned naturally with the soft-skin wheel with all your limbs unencumbered.

The starting key was one of Superman's crystal palace jobs, a rectangular block of resin that you inserted into the dashboard, held there and listened to the shrill whirring of the starter motor. Seconds later a deeper growl took over. It was one of the first models that required you to keep your foot on the brake while pressing the ignition, which flummoxed nosey parkers when they asked to sit inside and start her up. Neil told them it would only work if they said their name very loudly three times, while he dabbed the brake with his own toe.

I shifted the chunky chrome lever into first and eased out onto the track. Having been round Dunsfold a million times, I went full gangster with the gas and tore past the start line towards the right kink before the first corner. You had to be careful there because the road crowned and fell away as the corner broke left. The camera crews loved filming from a side road on the right – a guaranteed close-up if you overcooked it. I hit the brakes on the DBS and stopped too quickly, released them and made a note to try harder next time.

I cracked the throttle; the V12 opened, and the rasping blare of the exhaust was addictive as the revs grew to 6800. The flat-out left kink towards the long radius right at Chicago was always a test for wayward machinery that lacked downforce. There were no big wings on the DBS, but a cautionary lift wasn't needed. It leaned hard, found support and I carefully climbed onto the brakes with no adverse transfers.

Usually the brakes on a road car caught fire after three hot laps around the *Top Gear* track, so I thought I would give five a go and see what happened. The carbon ceramic brakes were basically indestructible, and I could brake late everywhere without mercy. It was time to switch off the traction control.

Wheelspin happens whenever you accelerate hard enough for the driven wheels to break traction with the road surface and start spinning faster than you're actually travelling. If you're being a caveman, you can spin the wheels at 100 mph while stationary until the tyres explode. It also puts stress through the drive train, from the clutch through the gears to the drive shafts. Purely for science, I cooked the gas harder out of the final corner and tipped the DBS into a hearty broadside in third gear. It was wonderfully controllable and threw up some smoke and dirt from the grass bank to entertain Neil's crowd, who were happier now with their tea and bacon sandwiches.

One thing I remembered about the badass Vantage was the crunchy gearbox. Maybe I could batter this one into submission. I came out of the fast left of Follow Through at 115 and sped towards Gambon, where all the celebs used to make like ballerinas and spin off into the weeds. It was a weird corner because the airstrip was wide enough to land a jumbo jet, but narrowed as you turned left into what was a lane for taxiing. The surface also changed to a lower-grade tarmac slick with avgas.

There were many ways to skin this particular cat, my preferred method being to brake as late as humanly possible in a straight line, release some foot pressure as I turned the wheel and take a trip across the grassy concrete at the centre point of the corner. Coming out the other side was always a roll of the dice.

To spice things up I made some vicious downchanges, which the gearbox didn't object to but did snap the rear sideways. To make this look deliberate I accelerated and drifted through to nods of approval from the bacon brigade.

There was no brake fade, not a hint of trouble from the gears, and I drove like an animal for ten laps until a cornering shimmy suggested the rear tyres were stripped down to their metal cords. We repeated this for the rest of the day, and though it was hard work, someone just had to do it.

Daniel Craig came down to help us get rid of some more tyres. His piercing blue eyes lit up when he saw the Aston, and I thought, *Here we go*. We had built an assault course marked by cones for throwing the DBS around and doing some burnouts. DC had fast reflexes, hardly surprising after all the muscle-shredding training he was doing in between the underwater knife-fighting, roof-jumping and stuff that Bonds do. Smoke poured in through the door seams as we toasted another set of rubber while circling Gary Powell's feet, then we did some fast laps of the track, and I gained some empathy with all the poor souls I had driven round there. DC kept his foot in at Follow Through and laughed as my white knuckles clamped around the Jesus handle. We returned to base, where he stepped out, straightened his jacket and said with a half-smile, 'Thank you, Ben. Always a pleasure.'

The poise of the DBS was superb, and nothing affected the attitude of the chassis except the wheel. That was unusual. With a front-engined car the hardest part to get right, or get rid of, is the sensation of being tied to the end of Eeyore's tail, with the engine weight at the front acting like the pin and the tail swinging round it. Ferrari made the best mid-engined sports cars in the world, but they struggled to make a front-engined tourer that didn't make you nervous. To support the engine's weight they stiffened the front suspension, which made cars like the California dart around in the dry and really misbehave on a wet surface.

The pinnacle of fear was provided by the Ferrari 599 GT0, which I had piloted around the same track a few months prior on *Top Gear*. It was a road-going version of the track-only FXX. The front end felt dead hard, and to complicate matters the suspension used magneto dampers that could think for themselves by effectively adjusting the viscosity of the liquid inside to make them hard or soft. Cornering was like walking across a floor covered in ball bearings in the dark. When I guessed wrong at Follow Through at 110 mph, I spent the next half a mile dealing with the consequences as it understeered, snapped sideways, understeered again and so on. Little wonder the test driver had shunted it a few weeks earlier . . .

My point being that if a firm like Ferrari found it hard to balance their GTs, you had to salute anyone who did it right – and the majestic DBS was a doddle by comparison.

Unit 20

Q had many incarnations in the worlds of Bond and Aston Martin. You couldn't swing a cat in the Cotswolds without hitting one of them, usually found lurking inside a hangar at the end of some backwater industrial estate. Astons' Q Branch was discreetly located at Unit 20, a large metal-roofed building that looked like all the others. Behind the tinted windows, however, spread the wings of the AM logo, and beyond that was where they kept their secrets.

Fraser Dunn was the Aston Martin chief engineer who oversaw their 'skunk works' projects, or as he put it, 'anything exotic or ultra-performance'. We had met in Garda on the set of *Quantum of Solace*.

These days any model name starting with the letter V meant Very Fast. The ultimate among these was the Vulcan. I met this £2 million machine in its natural habitat at a private members' club for the super-wealthy in 2017. Thermal Raceway is situated in the remote high desert near Palm Springs, California and has five miles of manicured tarmac with luxury storage facilities for fast toys. My mission was to demonstrate the qualities of Michelin's world-class tyres with the Vulcan, hang out with celebrities like Keanu Reeves and eat Michelin-star food all day long. It was gruelling. Michelin had assembled the finest chefs from around the world, including lord of fine beer, Daniel Burns out of Brooklyn, to launch their splendid new tyre compounds, and every dinner was six tantalizing courses of broths, mouth-watering foams, exquisite sauces, meat, fish . . .

One morning after, I waddled out to the track to assess the Vulcan. Having driven numerous over-powered nails in my time, such as the early Koenigseggs and every TVR ever made, I had wondered whether the Vulcan fell into the category of poorly executed nightmares. It was based on the limited-edition Aston One-77, which had previously escaped my attention. The One-77 looked similar to a modern DBS except for some pronounced shark gills and a miniature rear wing that popped out like a pole-dancer's undergarment. Hidden beneath this camouflage was a ground-breaking rigid carbon-fibre monocoque chassis and oodles of power from an enlarged version of the tuned V12 in the GT3 racer.

Q Branch had taken the One-77 monocoque chassis and used it to build an all-out track leviathan – not for racing, just for lulz. The Vulcan's 7-litre engine was good for over 800 bhp and sub-3-second acceleration to 60 mph. Full-scale aerodynamics included a vast rear wing, flat floor, gargantuan front splitter and gaping air scoop to create a full ton of downforce. It had the face of a prowling tiger, colossal tunnels exiting the broad bonnet, exhaust vents along the running board resembling those of a Spitfire and a rear end that contained an array of air-slicing dagger boards. This was a DBR9 with the safety catch off and the ultimate expression, visually at least, of beauty in a GT.

'I'd be careful in that,' warned one of the local instructors. Maybe he was right, but then he was driving a Boxster.

Fraser stood next to it trying to look unimpressed, with his arms folded and sunglasses on to block out the unfiltered rays bleaching our surroundings.

'How is it?' I asked.

'It's good.'

'Is it a science project or is this the real deal?'

He just laughed and opened the door, which swung on its hinge without a hint of deflection. The cabin was immaculate. No signs of the swarf, torn leather or washer screws on the seat that I associated with mega-powerful machines, which had a tendency to spray you in the face with petrol or for their wheels to fall off.

The steering wheel was a work of art. The bulbous centre section was made of carbon weave with the AM logo embossed. It extended out to curved handles with precisely placed control clusters. Many racey machines have buttons in all the wrong places and make you look an imbecile whenever you turn the wheel by activating the wipers, horn or lights. You could have plugged this wheel into the Starship *Enterprise* and dispensed with the crew. On the left a radio button handled comms; below was the full-beam headlight flasher for ordering paupers to piss off out of your way, while on the right you had a pit lane speed controller to placate the safety wombles and a neutral button to instantly put you out of gear, the faster to be handed refreshment in the pits.

Below these were a pair of significant dials. One adjusted the degree of anti-lock braking and the other handled traction control. The range went from Nanny holding your hand if you had a bad dream, to a night on the tiles with Attila the Hun when anything goes.

Fraser passed me my skid lid and I plugged in the comms. They weren't setting me loose without a means of recall. I pressed the start button and heard the whir of the starter motor deep inside the echo chamber. I played with the throttle. It responded instantly up and down the scales, creating a crisp bark. I pulled the paddle shift for first gear and eased the clutch, which was nothing like as brittle as the DBR9's.

With the instructor's words of caution ringing in my ears, I played it safe to start with and kept the 800 horses tethered by Nanny. With the wheel straight I gave it a gust of power, the Michelin slick tyres absorbing the energy and catapulting me forward. I applied the brake to see what it could do. The nose dipped with precision, and the Vulcan slowed without wavering from its line. The poise and resolve in the suspension was *perfect*.

I gave it a healthy belt of power coming out of a second-gear left and felt the restraint of the traction control as it prevented the tyres from stretching into their prime operating zone. The simplest way to gauge the Vulcan's true nature was to turn the damned system off, so I twisted the knob around to 0 and held on.

At full power the Vulcan sang an angelic chorus of trumpets, and you felt the downforce inching the suspension into the track. I reduced the ABS down to more of a background watchman and let rip with everything the Vulcan had to offer. On full brake the lightweight Brembo carbon-ceramic discs took every advantage from the aero grip, blotting away the speed in a way that outshone even the DBR9. It literally screeched to a halt as the raw racing pads clamped so noisily onto the discs they would get you an ASBO for noise pollution in civvy street.

The next corner was a double left, and on the exit I decided to feed the power in until something happened. The rear slipped predictably into oversteer, the tail yawed, and the tyres rattled the exit kerbs, putting the orchestra on notice to unleash a riotous uproar as the revs lashed up and down with the wheelspin. The power was so linear from the tiniest input of the throttle up to the maximum that I knew instantly that this was the finest and most full-blooded front-engined car I had ever driven.

I took some famous YouTubers for a series of passenger rides, and each motormouth turned into the same silent movie of gob-smacked fear, excitement and bowel control.

The seamless paddle shifting, revving to 7500 rpm, took me up to 175 mph on the straight, barrelling up behind the little Porsches and BMWs travelling so much slower that they looked parked. I secretly wanted them to get in the way to make use of the arrogant flash button, but they meekly pulled over to be buffeted by the wall of air rebounding off my aero. Never had a car matched the name of its spirit animal better than the Vulcan. The 80s jet bomber that had stirred my soul as a kid had been reborn. How could you possibly top that . . .?

Valkyrie

Valkyrie is Aston Martin's first ever hypercar and it leaves nothing in reserve.

Aston Martin

Adrian Newey had won more Formula 1 titles as a designer than any single driver. If you drove one of his cars well, regardless of the team, it was more than likely that you would win a Grand Prix. Before every season Newey scanned the rulebook of technical specifications to separate what the rules said from what he could get away with. Then he drew pretty pictures of what his car would look like, by hand.

School had been a chore for young Adrian except for occasional experiments with physics, such as shooting the teacher with a Bunsen-burner-powered pea-shooter. During the hippy craze, when the headmaster banned platform shoes and used a penny coin to measure the height of the boys' insteps, Newey's solution was to bridge the offending gap between the heel and toe with a piece of aluminium. His school reports would have impressed W. O. Bentley; he was labelled 'slapdash' and 'extremely silly' before he was politely asked to take his ideas elsewhere.

Some of his more superlative designs were ruled out of Formula 1 too when they were deemed so good that they were killing the sport, such as traction control and active suspension. Nothing was wasted, though; these ideas went into a vault and waited for the right occasion to present itself.

That moment arrived when Aston Martin approached Adrian and Red Bull Racing about designing the ultimate expression of a road car. This freed the mind of a man whose design approach could be summed up by everyone who had ever worked with him as uncompromising. He took the unwavering view that if the pinnacle of motoring was Formula 1, then the road car must have F1 performance.

Back at Unit 20, Fraser walked me towards the operating theatre, where a group of focused technicians examined the exposed parts of Aston Martin's future. The Valkyrie was stripped down to its core, like a captured stealth fighter revealing its secrets. The lean carbon-fibre monocoque was dripping with wiring looms; the wheels were off to reveal the complex hubs with streamlined cooling to the brakes, and the wishbone suspension looked like it was built for heavy duty.

Fraser's job was to ensure Aston Martin's practical objectives were met, such as building to a price that customers could actually afford. He also had the unenviable task of explaining government standards to the boffins in their F1 bubble. There were swathes of rules that had Newey reaching for the Bunsen burner, from crash and pedestrian safety systems to emissions controls and warranty considerations. One victory for both sides came in the battle over the required central high-mounted rear brake light, which interfered with Newey's rear wing profile. They solved this by procuring one of the world's most powerful LEDs the size of a pin head and installing it within a 5.5 mm light inside the spine of an air snorkel. It had to be that big in order to accommodate the minimum 5 mm required for the e-mark certification stamp.

To achieve F1 lap times required immense horsepower and fantastic aerodynamics. Fraser favoured the idea of doubling the 2.3-litre EcoBoost engine from a Ford Focus to generate an easy 1000 bhp, but that wasn't enough for Newey, so Cosworth was asked to come up with something 'mental'.

One option was stretching their IndyCar V8 into a V12, but Newey wanted a specific 65-degree V angle to marry with the package he envisaged, which meant creating a completely new unit. To meet emission limits they used a longer piston stroke to achieve a cleaner burn, as opposed to a shorter stroke with a broader piston head, and this grew the dimensions from 6.0 to 6.5

litres. The result was the most powerful naturally aspirated engine that has ever been built, and probably that ever will be. To boost power at lower rpm for pootling around city centres, they added a 160 bhp electric motor, taking the grand total to 1200 bhp.

The Valkyrie engine was a 'stressed component', which meant it formed part of the chassis structure itself, handling the extreme stresses of cornering and acceleration. To provide strength while keeping weight to a minimum, most of the motor's internal components were machined from solid, such as the titanium conrods, the forged F1-grade pistons and a crankshaft that takes six months to be whittled out of a solid steel bar. Weighing just 206 kilos, the V12 engine screamed furiously all the way to 11,100 rpm. You had every incentive to take it there because peak power was right up at 10,500. To help transfer its power to the pavement, torque vectoring divided the drive to each individual wheel according to the available grip.

Newey's genius for aerodynamics was wonderfully old-fashioned. While most designers now relied on computer-aided software to produce their drawings, Newey picked up a pencil in his left hand and drew the entire design from memory. Observers noted that his mental calculations rarely deviated by more than a single percentage point from the proven figures of the ultimate design. It's fair to say that he could view a wing profile and visualize the numbers of the corresponding flow dynamics in his head.

The finished article had an S-curvature from front to rear, and the car's original frontal area was a flat edge, so it didn't look much like an Aston. Fortunately for Marek Reichman, Astons' chief creative responsible for things like styling, Adrian owned a

DB4GT and was gently persuaded to accommodate the historic, aesthetically pleasing grille into the nose profile.

The Valkyrie weighed nearly 1100 kilos and produced 1800 kilos of downforce. The next mission was to control the delivery of these forces through their essential medium, the tyres. To maintain a constant pressure on its Michelin Cup 2 performance tyres, the Valkyrie used active aerodynamics, suspension and damping. This allowed it to adapt the aero forces being generated at different speeds to balance the handling characteristics, in real time, during cornering to position the car's weight and attitude in such a way as to provide maximum support and grip. G-forces would be high.

The body and chassis monocoque were one and the same, an interconnected aerofoil made from carbon fibre. The first example that rolled off the production line at Red Bull took 1000 man hours to construct and laminate, a tad laborious with 150 customer models to weave. Multimatic engineering was engaged to manage the production process, and adjusted the chassis tub and many of the installations into a more practical solution for making in numbers.

The Valkyrie is still in development but will undoubtedly blow the doors off the competition. It will be priced at a cool £2.25 million. Max Verstappen, a Formula 1 driver renowned for inspired overtaking moves and flamboyant car control, described the Valkyrie's handling as 'insane'. It may require an ejector seat for the less well endowed.

Perfect Is the Enemy of Good

The V8 Vantage line was resurrected by Ian Callum's pen as the smaller brother of the V12-powered DB cars. When it launched in 2005 it was a neat entry-level Aston with excellent proportions and a decent wallop of 380 bhp. The only problem was that it lacked the insanity required to slide it far enough up the scale of sex appeal that one might expect in an Aston.

I was returning from a *Top Gear* job on the cross-channel ferry in 2008 and noticed a Vantage driving off the exit ramp with conspicuous attention from its support team. By the time they had covered the unusual vents in the hood it was too late. I knew what they were up to. Six months later I was standing next to an early production model in a lay-by near Chipping Norton.

The passenger seat had been removed and the camera team were salivating over the mechanical slider tracks that had replaced it. The tracks were designed to slowly traverse a high-definition camera so that it could film the breathtaking scenery of the Welsh mountains and then pan slowly towards Clarkson for him to experience an epiphany to an emotive track by Brian Eno.

Astons had dropped their ferocious V12 into the lightweight, easy-going Vantage, and suddenly the girl next door was Gisele. Jeremy thought the V12 Vantage was three times wonderful and a monument to a bygone era. The PC brigade was waging war on fast cars, and even *Top Gear* was being targeted by the safety wombles, who wanted to ban it. Jezza feared this was the end of the line.

The project had begun in 2006 with Fraser's special projects team. They were unable to obtain official sign-off for their experiment so they just grabbed a spare Vantage chassis and threw a V12 in. Codenamed Esmerelda, the recalcitrant Gypsy Princess with a heart of gold, it was expected to 'understeer like a pig and have no traction', but the combination was an instant success and was formally commissioned for development.

I was filming the Vantage around the rolling Cotswold countryside along a passage of road I knew well, practically on first-name terms with the lawmen patrolling the district in their unmarked cars. The undulating curves of the roads were an idyllic test bed for the Vantage. Sure, the power was fabulous, as was the rigid shorter-wheelbase chassis. The problem was that the motor added 100 kilos to the front axle, which bounced around like a space hopper as a result. But it proved the concept of the smaller car with more power.

The ride and handling team whittled away at the suspension, adding stiffer springs and dampers as well as Bilstein adaptive dampers – another dreaded magneto system – but Aston Martin cleverly under-deployed the magneto control elements to avoid the feedback loop of Hail Marys I had experienced driving the Ferrari and other cars like the Corvette ZR1. The end result was a masterpiece: the V12 Vantage S. The weight was perfectly controlled; the power remained savage, and it had no vices. Formula 1 engineering genius Adrian Newey agreed, telling DJ and car collector Chris Evans it was 'simply the greatest road car ever made'.

In fact the S was so good it created a problem. It was better and faster than the latest, and more expensive, DB11.

20/20 Hindsight

The DB line was coming under pressure. The successor to the DBS was the Vanquish, costing just shy of £200,000. If you had a masochistic gene and kept a superbike in the garage to scratch the itch on a Sunday morning, then you might opt for the DBS Superleggera at £225,000. It boasted a garish open grille, the better for feeding air into the twin-turbocharged V12, which served up an angry dollop of torque (700 lb/ft) in the mid-range that crept up on you quietly and smacked you around the face if you weren't paying attention in third gear.

In 2013 Aston Martin signed a technical partnership with Mercedes-Benz's performance arm, AMG. This was primarily aimed at tapping into AMG's wealth of experience with turbocharging its V8 motors. A secondary benefit for Astons would be an upgrade from *Apollo 11*-style cabin instruments to something modern with Wi-Fi and stuff like satnav that actually worked.

For future models we bade farewell to the magnificent, gurgling V12 across most of the range and welcomed in AMG's 4-litre twin-turbocharged V8. The more compact motor could be positioned lower in the chassis to improve the centre of gravity, and further inboard to balance the weight distribution fore and aft, as well as giving the driver that sense of being central to the universe.

An all-new aluminium platform was crafted to make the most of the engine's compact proportions, and Aston Martin's design director Marek Reichman deployed his art with aplomb.

The full-length version was the DB11, priced at £155,000, a luxurious grand tourer with room in the back for small children while your partner admired the view of the scenery up front. It might have stolen the show had it not been for its little brother. The shorter Vantage was like a mini-Vulcan: pugnacious but sophisticated and with looks that could kill. At 1530 kilos, it weighed 170 kilos less than the DB11 and with the same 503 bhp propelled you to 60 mph in 3.6 seconds. Handling was so good that when Silverstone bought a fleet for their track days, they didn't even modify them. You could drive it like you hated it and it would ask for more. Cornering grip was off the charts, but the car's attitude remained fully alive to driver input. This would have been Tadek Marek's modern car of choice for that rush across the border.

At £120,000 the Vantage was Astons' cheapest model, and from where I was sitting it was a no-brainer in terms of value and performance. Complaints about its lack of rear seats could easily be resolved by sending the family separately via public transport. I was surprised to learn that it used an electronic rear differential because my heavy foot usually overheated those and the Vantage never skipped a beat. The eight-speed automatic shift was stellar too, but . . . they recently launched a manual version with a mechanical diff and that really talks to my inner caveman. Yet again I find myself in love with the Vantage.

Arriving late for the party is the DBX, Astons' first sports utility vehicle. The benefit of being fashionably late is being able to observe the best and worst features of those already present. Aston Martin benchmarked a formidable range – Range Rover Sport SVR, BMW X6M, Porsche Cayenne Turbo, Audi SQ7,

Bentley – and they even deigned to take a spin in the chav-tastic Lamborghini Urus . . . While the outcome is hotly anticipated, Matt Becker, Astons' ride and handling guru, has already logged some impressive figures. The DBX pulls 1.25 G of cornering force, equivalent to the Vantage, and 1.4 G under braking, which matches the Superleggera – quite enough to shake and stir the entire family.

To fund this expansion, Aston Martin took a leaf out of Ferrari's book by taking the firm public to become Aston Martin Lagonda Global Holdings plc. Ferrari floated in 2015 and was initially valued at $10 billion. The share price investors paid nearly halved in the course of the following year, before rocketing in value, making the company worth an estimated $30 billion as of 2020. So far Aston Martin has won the race to the bottom, tanking from an IPO value of £19 per share in 2018 to a recent low of 27 pence, buying you half a Mars bar. If Ferrari's roller-coaster example can be duplicated, then hopefully investors can expect a stellar recovery on the way back up . . .

Count Louis Zborowski admired Aston Martins for their supreme handling capabilities but he was drawn to the abundant power of Mercedes. He would undoubtedly have approved of the technical partnership between the two companies – and been enchanted by the next turn of events.

Tobias Moers is an engineer who rose to command AMG, deploying a masterplan that quadrupled sales of its sports cars in seven years. Tobias is no pussycat. He is a fixer and in 2020 took the helm of Aston Martin Lagonda as CEO. Judging by the investors now aligning with the firm, his timing is superb.

Mercedes Benz owns 5 per cent of Astons, Merc's F1 team boss Toto Wolff recently bought 5 per cent for himself, and billionaire investor Lawrence Stroll owns 30 per cent of the equity. Just as the rules are changing in Formula 1 to cap spending and give the smaller teams a chance, Lawrence is taking Aston Martin into F1 powered by the same Mercedes engines that have dominated the series for six consecutive seasons. Astons' outgoing chief exec Andy Palmer once said, 'We don't make cars, we make dreams.' Nearly a hundred years after Aston Martin first competed in a Grand Prix, this dream might just come true.

DB5 Mark II

Tadek Marek's experimental V8 DB5 had shown the way back in the 1960s, and for many the DB5 is *the* iconic expression of Aston Martin, so it was always going to deserve another chapter.

Since we had filmed the final scene of *Spectre* where 007 drove away from a defeated Blofeld, things had developed apace and to keep up with the script of *No Time To Die,* an all-new DB5 was required to handle it.

Chris Corbould's special effects department linked arms with the special projects team at Aston Martin to build a stunt version of the DB5. In some cases movie cars are thrown together with cynical minimalism, the attitude being 'it moves doesn't it, so drive it.' That wasn't the case here.

The original DB5 weighed 1500 kilos, whereas the lean stunt machine weighed just over half that. Astons installed a 3.2-litre straight-six engine that developed around 333 bhp – an increase of 50 bhp which was directed to the rear wheels with a limited slip differential, making it a cheeky little varmint. The finished trim was so good that to look at, it was indiscernible from an original DB5: everything from the spoked wheels to the stunning chrome fenders and a body that shone like liquid mercury.

Opening the door took you by surprise because it was paper light. The hard bucket seat only came up as far as your armpits to mask it from the camera, and it was mounted onto a bare chassis since there was no need for a smart carpeted interior. The hydraulic handbrake mounted in the centre console was shorter than usual but a firm tug snatched the rear tyres to pull a 180 or skid into a corner, while the pliable power kept you there. The gorgeous three-spoke wooden steering wheel was just like the original, and perfect for manhandling the silver bullet on its next mission.

No Time To Die

Matera in the Basilicata region, Southern Italy, was the site of one of the world's most ancient cave habitations with estimates of people living there for 10,000 years or more. Residents lived in a network of caverns carved into the soft limestone that crawled up the top side of a deep valley, fusing the small dwellings with the pale mountain rock. The large stones used to pave the streets were worn like old leather by the infinite steps of civilizations past and present.

The modern-day visitor was greeted by hidden treasures at every corner, from the thirteenth-century Cattedrale de Matera at the highest point of the city to the morbid Church of Purgatory with its ghoulish reminders of mortality on its carvings and statues. At night the whole place ignited with mellow lighting that drew you into exploring the warren of paths through derelict districts, pulsing bars, grotto restaurants and an assortment of gelaterias.

Declared a UNESCO World Heritage Monument in the early 1990s, Matera was the perfect place to get away from it all, slow your pace and re-connect with a sense of timeless humanity. Yes, it was the perfect holiday destination.

'*Drone's going up!*'

The sound of angry hornets was the drone lifting off, its five helicopter blades lifting the camera into position above the back street. The voice of Dom Fysh was tailor-made for overseeing the second unit on an action movie, being clear, and even more importantly given the hardware we'd brought with us, loud. His role as first AD was to wrangle the crew and help the director shoot the scenes in a timely fashion. Line of sight in Matera was limited by the innumerable buildings upon buildings within walls and side streets going in every direction. To aid communication he had the loudhailer to end all loudhailers, his being wired to a networked array of large speakers. I walked past one as he cleared his throat and it nearly blew my head off.

Rally driver Mark Higgins was driving this section in the DB5, chased by the 'bad guys' in a pair of Jaguar XFs as well as some lunatics on bikes. My role was assisting Lee, the stunt co-ordinator, to jump in and drive whenever and in whatever was needed. This ranged from the occasional jaunt in the DB5 right down to the lowly Fiat Cinquecento.

Shooting for the action sequence began with Bond blasting down an alley towards an open piazza and a few weeks earlier, this square had revealed a major problem. Matera's smooth limestone bricks were so slippery that the vehicles could only totter around like Bambi on ice. There was nothing that could be done with the cars to speed them up and Lee was faced with the ultimate nightmare: a slow-motion car chase. The production team had a brainwave and sprayed fizzy cola on the surface . . . The magic ingredient acted like a glue and the tyres gripped the surface.

The next section along the Via Madonna Delle Virtu was the fastest run in Matera and you could feel the rumble of the bricks as the tyres vibrated across them. The road was just wide enough for two cars to pass side-by-side and was bordered by an abrupt kerb so it was messy if you ran wide.

Lee talked everyone through the action over the radio. Mark led in the DB5, chased by BMX legend Ali Whitton on a Ducati Scrambler, with Greek ace Evangelos in the first Jag and Martin the Russian at the back in the other one. Martin was the polar opposite of a yes man, but we loved him.

Each man was supposed to reply 'copy' on the radio to confirm he understood, otherwise the channel was kept absolutely clear for important transmissions.

The unit base was at the top of the Via Madonna in a dusty car park overlooking the gorge. We filled it easily. There were six DB5 stunt cars parked in a row and one of the two converted 'pods', plus two hero cars. The pod was a caged enclosure fitted to the roof with a steering wheel and pedals so that it could be driven from upstairs while the actors performed their scenes inside the car. Then there was the fleet of trucks for the grips, camera department, sound and most importantly the craft service wagon serving all manner of snacks and enough caffeine to put a spring in everyone's step.

In the middle of the jungle of alleyways, key grip Dave Cross was making some fine adjustments to the camera rig on the front of a Jaguar with his lump hammer. The front grille and nose panels were removed, replaced by a complicated frame of scaffold tubes and ratchet straps that attached to the chassis. The camera mount would sit about six inches from the floor to film the combined effect of ground rush in the lower part of the frame with the action filling the rest. Dave was the maestro when it came to managing ground clearances on these rigs. His narrow eyes hardened for a final assessment and he took a long draw on his fag while the cogs turned.

'Yeah, tell Lee you can send it, son. If you 'ear it touchin' that's just the skid plate.'

Lee had done a number of 'clean' runs with the heroes and now it was time to insert the camera car. I climbed aboard the Jag and replaced Martin's spot in the high-speed queue. The camera was invisible from behind the wheel so I pulled up to the rear wheel of Ali's Ducati to make a visual reference of the point where it would touch. His knobbly tyre had a contact patch with the road that was smaller than an egg so the last thing I wanted to do was brush it with the Jag's nose.

We lined up to run a half-speed rehearsal. I started on the DB5's tail with Ali on my right and we gunned it towards a cresting left that tightened on the exit. Even at half speed the tyres felt vague on the surface, like driving on putty. Ali switched across left and ripped it up a set of stone steps to a small landing area where Lee was running the show. The rest of us ran downhill, floated through a right-left curve and then slowed. These guys were well in the zone.

We re-set and as they tinkered with the camera I went over to Mark in the DB5. Underneath the smart tan jacket, the sleeves of his blue shirt had been hacked off for cooling so he looked a bit like Barney Rubble, and he was in full song crooning to some old Bing Crosby tune.

Ali shot past on the Ducati and skidded it around 180 degrees on the brake; he was living the dream.

'Evangelos,' I asked him since he was the most sensible one, 'was that half speed or did his foot slip?' pointing at Bing Crosby. It was worth asking because the cola-fuelled grip was changing all the time.

'That was half. It feels strange but yes, they can go faster on this than you think.'

The next run was full speed, the DB5 lighting up its rear wheels less than a car length ahead of me, with Ali close on my right fender. A grey helicopter appeared low over the canyon with its football-shaped, gyro-stabilized camera ogling the scene on a tight angle.

We compressed into a tight formation for the left, Ali made his move and shaved the front of the lens with his back tyre. I suppressed the road environment instinct screaming for me to slam on the brakes to avoid his rear wheel. Here my job was to accelerate through Ali to catch Mark as he laid on a graceful slide.

The chrome tail of the DB5 skimmed wide until I couldn't read the number plate, showing the left flank with the front tyre extended out of the wheel arch in pitch-perfect control. My Jag bludgeoned through the curve with the front tyres squealing and I knew Evangelos would be on my tail. We raced through the left-right curves and the mysterious surface behaved itself.

The pièce de résistance for the DB5 was a hotly anticipated scene where Bond seemed to have no escape. Chris Corbould went 'full combat mode' on the gadgets. He came up with mini bomblets that thinned down the opposition, Martin driving one of the machines that was blown up.

When Bond was surrounded in a tight square by henchmen blazing away with machine guns, he unleashed Q's winter project: a pair of large six-barrelled rotating machine guns extended from the headlamps. Daniel Craig side-slipped the clutch in first gear and spun the car in a tight 360 donut, hosing down the bad guys. Rather than using a blank-firing replica that would jam for lack of back pressure, Chris built his own mechanical guns with CGI providing the flash and smoke effect. A class touch was having the stream of spent brass casings spill out of the side vents behind the wheel arches.

No Time To Die was a remarkable showcase for Aston Martin, with four different models appearing in different scenes. We flew to Norway to shoot on the stunning Atlantic Ocean Road that snaked its way along the archipelago. I held a major soft spot for the V8 Not-Vantage from *Living Daylights* and on arrival discovered three of them lined up, fully re-conditioned and ready for service.

I climbed into the driver's seat with reverence. The all-black wheel was smaller than the old wooden number, but thicker and with stronger spokes. The dashboard was decorated with polished walnut and hard-wearing black leather. The only soft thing about the V8 was the expansive chair, which positioned the driver quite high in relation to the roof. I gunned the engine and felt the torque twisting the chassis on its springs.

'Mind yourself with that, alright, and keep it on the island,' said Lee from the door.

People kept telling me to be careful but coming from him it was unusual. The last time I saw Lee in action was filming *Jack Ryan*. He was chasing me on a Harley-Davidson when I slid a van across the width of a narrow tunnel, leaving him nowhere to go. He dropped the bike, skidded along the ground and slammed into the side of the van, which was exactly what the director wanted.

The helicopter took off and I fired up the V8, which chugged along while the costume department checked my collar was in the right location. The heli would be coming in hot across the water from the right side and aim to join me as the road climbed dramatically up from the ocean, like a maritime version of Eau Rouge from Spa, before dropping down the other side.

We set off and the pull of the 5.3-litre motor was strong as I fed in the 390 hp. The suspension absorbed the power easily, rather than squatting flat on its backside like a lot of the old muscle cars from that era. Nor was it too sharp. There was enough corresponding movement from the chassis to my inputs. I took her up to 95 mph in fourth and felt the heli floating into my peripheral vision, without ruining the shot by looking over my shoulder.

The wheel tightened in my hands as the road gained altitude and began to arc left. The Aston leaned easily into the springs, and I crested the blind rise as the chopper grew large alongside and married to my speed. I just had to ease off on the way down the steep drop to avoid accelerating out of the frame, but I could feel the car asking for much more. The acceleration at higher revs was seductive, and as soon as the heli veered away I buried the throttle to set it free for a few glorious moments, diving down towards the ocean and closer to Bond's final destination. I took a very long last look at the V8 Volante-not-a-Vantage on completion of duty. I shouldn't pick favourites but out of the Bond originals, it stole my vote.

Belonging

The genetic contest raging within Aston Martin will continue to purify the breed. With the postmodern DB line producing timeless winners, perhaps we can afford to enjoy a little luxury at the expense of ultimate performance. We can leave the bare-knuckle brawling to the latest line of superlative V-cars.

If driving Astons over the years has taught me anything, it is to stop thinking that *this is as good as it gets*, because they always find more. The way they get ahead of themselves is what makes their story so endearing, regardless of how difficult it has made running a business. Nobody sees them coming. Then, out of nowhere, Lionel Martin shreds all your Brooklands records, Zborowski appears hot on your tail during a Grand Prix, Moss annihilates your Sportscar domination, the Bulldog defines what a hypercar is, James Bond knocks you off the road and, well, others start to follow their lead.

We live in unusual times, when the personal touch is often outsourced or digitized, and quality is ignored by bean-counters in favour of the bottom line.

It doesn't have to be.

Recently, I took a trip to Aston Martin's works factory at Newport Pagnell. Old-school craftsmanship has been fused with cutting-edge technology. Men were turning the body panels, beating the curves by hand and spinning parts with lathes in conjunction with sophisticated CAD-designed renderings, to produce what appeared to be flawless reproductions of vintage models.

Among other projects, the works team is currently hand-building twenty-five *Goldfinger*-spec DB5s for what they call the 'continuation series', fully loaded with Q's gadgets and commanding a price tag of £2.75 million each. They won't be road legal as they roll off the line, but you can have them homologated after purchase and then blaze merrily away at the traffic with the twin Browning machine guns.

The tailor-made operation at the works factory is surprisingly forward-looking, even as the world moves towards electrification. President of Aston Martin Works Paul Spires showed me a glimpse of the opportunities ahead for vintage car lovers with a DB4 converted seamlessly to an electric motor. Electric torque through an old-school drive train was a beautiful marriage. You kept all the feel and feedback coming through the mechanical steering rack, while the extra power livened up an ageing design.

I looked out through the showroom window towards a construction site on the other side of Tickford Street. A lone Tudor-style cottage marked FOR SALE stood defiantly in the centre, at least for now. It was David Brown's old office at Sunnyside, its disposal a stark daily reminder of the consequences of evolution. But such is the way with this prince of British car makers. We can feel free to reminisce, but Aston Martin's soul will always belong to the future.

BIBLIOGRAPHY

I couldn't have written this book without these superbly detailed works. Other than the interviews that I conducted personally, all the quotations I have used came from these excellent texts and I thank the authors for my edification.

Bentley, W. O., *The Autobiography of W. O. Bentley*, Hutchinson of London, 1958.

Bowler, Michael, *Aston Martin: The Legend*, Parragon, Bristol, 1998.

Courtney, Geoff, *The Power Behind Aston Martin*, G. T. Foulis & Co. Ltd, Oxford, 1978. [*Interviews with Bertelli, David Brown and key designers.*]

Davis, Sammy, *Motor Racing*, Iliffe & Sons, London, 1932.

Davis, Sammy, *My Lifetime in Motorsport*, Herridge & Sons, Beaworthy, Devon, 2007.

Demaus, A. B., *Lionel Martin: A Biography*, Transport Bookman Publications, Middlesex, 1980.

Donnelly, Desmond, *David Brown's: The Story of a Family Business*, Collins, London, 1960.

Dowsey, David, *Aston Martin: Power, Beauty and Soul*, Images Publishing Group, Victoria, Australia, 2010.

Edwards, Robert, *Stirling Moss*, Cassell and Co., London, 2001. [*A wonderfully detailed authorised biography.*]

evo, *Aston Martin: Behind the Wheel of a Motoring Icon*, Mitchell Beazley, Hachette UK, London, 2017.

Fleming, Ian, *Goldfinger*, Jonathan Cape, London, 1959.

Gauld, Graham, *Cliff Allison: From the Fells to Ferrari*, Veloce, 2008.

Gibbard, Stuart, *The David Brown Tractor Story*, Old Pond Publishing, Ipswich, 2003. [*A richly researched story of DB's main businesses and contribution during WW2.*]

Hunter, Inman, Ellis, F. E. & Coram, Dudley, *Aston Martin: The Story of a Sports Car*, Motor Racing Publications, London, 1957.

Jarrott, Charles, *Ten Years of Motors and Motor Racing*, G. T. Foulis & Co. Ltd, London, 1906.

Macintyre, Ben, *For Your Eyes Only: Ian Fleming and James Bond*, Bloomsbury Publishing, London, 2008.

Macintyre, Ben, *Operation Mincemeat*, Bloomsbury, London, 2010.

Montagu, Ewen, Introduction by Stripp, Alan, *The Man Who Never Was*, Oxford University Press, Oxford, 1996.

Moss, Stirling, *A Turn at the Wheel*, William Kimber, London, 1961.

Moss, Stirling, *Le Mans '59*, Cassell and Co., London, 1959. [*A ripping yarn by Sir Stirling, told from the heart and at the wheel.*]

Moss, Stirling & Taylor, Simon, *Stirling Moss: My Racing Life*, Evro Publishing, Sherborne, 2015.

Newey, Adrian, *How To Build A Car*, HarperCollins, London, 2017. [*Essential, hilarious and inadvertently educational.*]

Nixon, Chris, *Racing with the David Brown Aston Martins*, Volumes 1 & 2, Transport Bookman, Middlesex, 1980. [*The inside scoop on the quixotic road to victory and the brazen personalities who did it. Great fun.*]

Nixon, Chris, *Sportscar Heaven*, Transport Bookman, Middlesex, 2002.

Pirie, Valerie, *Ciao Stirling*, Biteback Publishing, London, 2019. [*Behind this great man were some great women, and one of them wrote a great book. Meet "Viper".*]

Rankin, Nicholas, *Ian Fleming's Commandos: The Story of 30 Assault Unit in WWII*, Faber and Faber, London, 2011.

Shelby, Carroll, *The Carroll Shelby Story*, Graymalkin Media, USA, 2020.

West, Nigel, *MI5*, Triad Granada, Reading, 1981.

Wilson, David, *The Racing Zborowskis*, Vintage Sports-Car Club, 2002. [*An absolute gem; the life and times of Count Louis and friends.*]

Wyer, John, *The Certain Sound*, Automobile Year, Lausanne, Switzerland, 1981. [*John's wry account of an epic racing life, from Astons through to Ford and his exploits with Gulf.*]

Yates, Brock, *Enzo Ferrari: The Man and the Machine*, Penguin Books, London, 2019.

Periodicals

Car Magazine

Aston Martin Owners Club Magazine

Aston Martin Heritage Trust Magazine

Light car

Autocar

Motor

Motor Sport

Sources

Clifton College archives / school reports

Aston Martin Heritage Trust archives

The Lagonda Club Archives

SPECIAL THANKS

To everyone at Aston Martin for helping me to assemble these stories, as well as giving me the opportunity to express myself behind the wheel of your wonderful speed machines over the years.

It all began with my superb editor, Jon Butler, spotting the driving force behind Aston Martin and bestowing me with the honour of writing about it. My first stop was a visit to Aston Martin works' division to see some old friends, Julian Wren and Mel King, who kindly invited me to see the living history of the production line and get to grips with the ethos of the company with Steve Waddingham, the product specialist and unofficial historian of AM with an encyclopaedic knowledge of the marque.

From there it was off to the races to interview as many key players as I could. I thank them for 'putting me in the room' inside their world and for more than a little laughter: Nick Fry, Ian Callum, Marek Reichman, Fraser Dunn, Adam Brown, Stuart Gibbard, Maitland Cook, Matt Becker, Chris Corbould, Neil Layton, Lee Morrison, Arnold Davey, Ray Mallock, Darren Turner and David Brabham.

Without a doubt my close relationship with Astons has been largely forged thanks to the moviemakers and storytellers at EON Productions, who continue to delight audiences with the adventures of 007. I'm grateful to Stephanie Wenborn and Jenni McMurrie for overseeing this project, to Barbara Broccoli for the perpetual motion behind James Bond, and to DC for being the man.

My thanks to Corinne Turner at Ian Fleming Publications for access to letters that shed further light on Bond's DNA and his appetite for a fast motor; to the Aston Martin Heritage Trust for throwing open the doors to their museum and letting me dig into their meticulous records, notably to Rob Smith and John Wood for their time locating key files and photos; Ian Girling too, for your insights on Jock; Dirk Ude and Team DHL for the busy nights in "Shanghai"; Andrzej Ślązak, for sharing your research on the indomitable Tadek Marek; Transport Bookman Publications for invaluable source material by Chris Nixon; Lady Moss for kindly allowing me to quote Sir Stirling in his own words and Robert Blakemore at Ecurie Bertelli for braving the magic round-about with me.

Thank you, Mark Lucas, for your pillars of wisdom; Milly and Bethan, for wrangling the latest marketing innovations to bring this book into the world.

To Georgie and the Stiglets – thank you for all the tea, cake and forebearance.

Finally, thank you for reading this book – it's been a pleasure to share this impetuous journey with you and I hope you'll soon get your hands on the merchandise.